T0122360

Efficient Design of Variation-Resilient Ultra-Low Energy Digital Processors

Hans Reyserhove • Wim Dehaene

Efficient Design of Variation-Resilient Ultra-Low Energy Digital Processors

Hans Reyserhove
ESAT—MICAS
KU Leuven
Seattle, USA

Wim Dehaene
KU Leuven
Heverlee, Belgium

ISBN 978-3-030-12487-8 ISBN 978-3-030-12485-4 (eBook)
https://doi.org/10.1007/978-3-030-12485-4

Library of Congress Control Number: 2019931385

This Springer imprint is published by the registered company Springer Nature Switzerland AG.
The registered company address is: Gewerbestrasse 11, 6330 Cham, Switzerland

Preface

This book describes the continuation of the work in the KU Leuven-MICAS division on ultra-low energy processors. In previous work (Reynders N., Dehaene W., *Ultra-Low-Voltage Design of Energy-Efficient Digital Circuits*, Springer, 2015), we showed that using transmission gates can result in variation-resilient energy-efficient digital signal processing blocks. However, at that time, these techniques could only be used in a handcrafted way on relatively regular data paths. Besides further optimizing the circuit techniques, we take this technique a step further in this work. Our transmission gates end up in a library that is compatible with regular digital design flows. An extension to the flow, to deal with the differential nature of the transmission gate-based logic, is also described. This results in an ARM Cortex-M0 as a demonstrator of the excellent energy efficiency these techniques allow.

The circuits presented run at a supply voltage below 500 mV. This calls for large design margins, even if intra-die variability is properly dealt with. These margins are canceling part of the energy improvements that comes with low power supply voltages. To deal with this, we introduce in situ timing detection in the system. Late transitions on paths with small timing slack are detected, and by means of special soft-edge flip-flops, timing errors are avoided. This results again in an ARM Cortex-M0 which can now be operated at very low energy without the need for large margin on the power supply.

This book is the result of 5 years of PhD work: a close cooperation between a young researcher and his advisor. As you will read, it was a very fruitful cooperation which we enjoyed a lot. We hope that sharing our results with you also brings you the professional achievements you strive for.

Seattle, USA
Heverlee, Belgium
December 2018

Hans Reyserhove
Wim Dehaene

Contents

Abbreviations and Symbols

AHB	Advanced high-performance bus
ALU	Arithmetic logic unit
AMBA	Advanced microcontroller bus architecture
APB	Advanced peripheral bus
BTWC	Better-than-worst-case
CCS	Composite current source model
CMOS	Complementary metal-oxide-semiconductor
CPF	Common Power Format
CSCD	Current-sensing completion detection
CTS	Clock tree synthesis
DIBL	Drain-induced barrier lowering
DMIPS	Dhrystone million instructions per second
DS	Double sampling
DSP	Digital signal processing
DVFS	Dynamic voltage frequency scaling
DVS	Dynamic voltage scaling
DW	Detection window
ECG	Electrocardiogram
ECSM	Effective current source model
EDA	Electronic design automation
EDAC	Error detection and correction
EDP	Energy-delay product
FD-SOI	Fully depleted silicon-on-insulator
FF	Fast nMOS fast pMOS process corner
FS	Fast nMOS slow pMOS process corner
FTDI	Future Technology Devices International
FO4	Fan-out 4
GALS	Globally asynchronous locally synchronous
GP	General purpose technology type
GPIO	General purpose input/output signal
HDL	Hardware description language

HVT	High threshold voltage transistor
I/O	Input/output
IoT	Internet-of-Things
IP	Intellectual property
IRQ	Interrupt request
ISR	Interrupt service routine
JTAG	Joint Test Action Group
LEF	Library exchange format
LVT	Low threshold voltage transistor
LP	Low power technology type
μP	Microprocessor
MC	Monte Carlo analysis
MDP	Minimum delay point
MEP	Minimum energy point
MMMC	Multi-mode multi-corner
MOSFET	Metal-oxide-semiconductor field-effect transistor
MTBPF	Mean time between potential failures
NFC	Near-field communication
NLDM	Non-linear delay model
NMI	Non-maskable interrupt
nMOS	n-channel MOS transistor
NoC	Network-on-chip
NVIC	Nested vectored interrupt controller
OCV	On-chip variation
PCB	Printed circuit board
pMOS	P-channel MOS transistor
PMU	Power management unit
PoFF	Point-of-first-failure
PVT	Process-voltage-temperature condition
PWM	Pulse width modulation
RDF	Random dopant fluctuations
RFID	Radio-frequency identification
ROM	Read-only memory
RSCE	Reverse short channel effect
RTL	Register-transfer language
RVT	Regular threshold voltage transistor
SBOCV	Stage-based on-chip variation
SDC	Synopsys design constraints
SDL	Set-dominant latch
SF	Slow nMOS Fast pMOS process corner
SoC	System-on-chip
SOI	Silicon-on-insulator
SRAM	Static random access memory
SS	Slow nMOS Slow pMOS process corner
SSTA	Statistical static timing analysis

TD	Transition detection
TG	Transmission gate logic gate
TT	Typical nMOS typical pMOS process corner
TTM	Time-to-manufacturing
UART	Universal asynchronous receiver-transmitter
ULV	Ultra-low voltage operation
UTBB	Ultra-thin buried box
UVFR	Unified voltage frequency regulation
VCO	Voltage-controlled oscillator
VLSI	Very-large-scale-integration
VR	Virtual reality
WIC	Wake-up interrupt controller

Symbols

C	Capacitance
C_{total}	Total capacitance
E	Energy
E_{dynamic}	Dynamic energy
E_{margined}	Margined energy
E_{optimal}	Optimal energy
E_{PoFF}	Point-of-first-failure energy
E_{static}	Static energy
E_{tot}	Total energy
ep_i	Number of endpoints in bin i
f_{clk}	Clock frequency
f_{max}	Maximum clock frequency
f_{target}	Target clock frequency
FF_i	Flip-flop in stage i
I_{d}	Drain current of a transistor
I_{ds}	Drain-source current of a transistor
I_{leak}	Leakage current
I_{off}	Off-current or leakage current of a transistor
$I_{\text{off},n}$	Off-current of the nMOS transistor
$I_{\text{off},p}$	Off-current of the pMOS transistor
I_{on}	On-current or drive current of a transistor
$I_{\text{on},n}$	On-current of the nMOS transistor
$I_{\text{on},p}$	On-current of the pMOS transistor
L_{gate}	Length of the gate of the transistor
n	Process-dependent parameter
$NM_{\text{L/H}}$	Low/high noise margin
P_{dynamic}	Dynamic power
P_{leak}	Leakage or static power

P_{static}	Static or leakage power
P_{tot}	Total power
p_i	Non-EDAC monitored logic path
q_i	EDAC monitored logic path
RNM	Relative noise margin
R_{on}	On-resistance of a (combination of) transistors
S	Subthreshold slope
t_{borrow}	Time borrowing amount
T_{clk}	Clock period
T_{system}	System clock period
$t_{\text{clk-q}}$	Clock-to-output delay
$t_{c,\text{clk-q}}$	Contamination clock-to-output delay
t_{d-q}	Input-to-output delay
t_{DW}	Detection window delay
t_{hold}	Hold time
$t_{p,\text{logic},i}$	Propagation delay of logic stage *i*
t_{prop}	Propagation delay
t_{setup}	Setup time
t_{slack}	Slack delay
V_{BB}	Body voltage of a transistor
V_{dd}	Supply voltage
$V_{\text{dd,MEP}}$	Minimum energy point supply voltage
$V_{\text{dd,margined}}$	Margined supply voltage
$V_{\text{dd,min}}$	Minimum supply voltage
$V_{\text{dd,nom}}$	Nominal supply voltage
$V_{\text{dd,PoFF}}$	Point-of-first-failure supply voltage
$V_{\text{dd,step}}$	Supply voltage step size
V_{ds}	Drain-source voltage of a transistor
V_{gs}	Gate-source voltage of a transistor
$V_{I,L/H}$	Low/high input voltage
V_{IO}	Input/output domain supply voltage
$V_{O,L/H}$	Low/high output voltage
V_{sb}	Source-bulk voltage of a transistor
V_{sd}	Source-drain voltage of a transistor
V_{sg}	Source-gate voltage of a transistor
V_{ss}	Ground voltage
V_T	Threshold voltage of a transistor
V_{T0}	Threshold voltage of a transistor for V_{sb}=0
W_{pMOS}	Width of the gate of the pMOS transistor

α	Activity factor
γ	Body effect coefficient
Δ	Delay or difference
η	DIBL coefficient

μ_x	Mean value of x
σ_x	Standard deviation of x
Ψ	Cumulative distribution function
ψ_0	Surface potential

List of Figures

List of Tables

Chapter 1
Energy-Efficient Processors: Challenges and Solutions

Abstract This work tries to combine the described challenges that energy-efficient microcontrollers face in sub-micron CMOS technologies. An ultra-low energy consumption with fast enough performance while being variation-resilient is the triple combination this work targets. Ideally, this system is realized using an efficient design process that helps the designer to improve the system as much as possible. To accomplish this, variation-resilient building blocks and design techniques to operate at ultra-low voltage are presented. This results in the efficient implementation of several microcontroller prototypes fabricated in 40 nm CMOS technology that achieve state-of-the-art performance and ultra-low energy consumption. This chapter gives a short overview of the main challenges faced in this work, and thereby introduces the remaining chapters. First, the concept of minimum energy operation is introduced. It is the main target for everything developed in this work. Second, CMOS technology and the problems it faces when operated for minimum energy are discussed. Third, efficient design through a standard cell based very-large-scale-integration (VLSI) design flow is elaborated on. Variations and their influence are the fourth topic. Fifth, current and possible future applications for microcontrollers and other systems that benefit from minimum energy operation and variation-resilient design are discussed. These topics set out the main challenges this work faces. Finally, the goals this work aims for are defined.

1.1 Introduction

The microcontroller is a key component in embedded devices today. All around us, modern day appliances incorporate a microcontroller as the main computing unit. Your everyday washing machine, the stylish camera-equipped drone, the lifestyle-improving activity tracker or the battery-constrained hearing aid. They all embed this programmable building block on their motherboard. It runs a piece of software to operate the control algorithm, read out sensors and actuate other components.

Over the last decades, microcontrollers have grown from 4-bit processing with a couple of bytes memory and a few input–output lines to the 32-bit processing units with large on-chip memory hierarchy and mixed-signal peripheral functions they

H. Reyserhove, W. Dehaene, *Efficient Design of Variation-Resilient Ultra-Low Energy Digital Processors*, https://doi.org/10.1007/978-3-030-12485-4_1

are today. The result is a jack-of-all-trades system ready for massive deployment in almost any embedded device. This is being done more than ever: [8] predicts a 2018 microcontroller market of 25 billion units worth close to 20 billion USD. This puts the average selling price per unit below 1 USD. For markets of such a scale, CMOS technology is the way to go. It allows cost-efficient fabrication of single-chip microcontroller implementations. Moreover, the advancements in CMOS technology increase the processing speed or improve the energy efficiency of these systems.

The microcontroller, being the workhorse of all these applications, definitely has a power impact. Especially when considering battery-operated devices that require a very long lifetime, the lowest possible energy consumption is mandatory [9]. Biomedical implants, wireless sensor nodes and RFID tags are applications that immediately come to mind [1, 11, 32]. Apart from the improvements brought by smaller feature size transistors, operating at a lower supply voltage can drastically improve the energy consumption of the microcontroller. But as the voltage and energy goes down, so does the processing speed. Achieving a fast enough operating speed for a variety of applications while minimizing the energy consumption through low voltage operation is one of the main challenges here.

The properties of transistors vary as a result of their fabrication process, their environment and their age. As transistor sizes shrink, the impact of these variations grows. Especially process variations have a large effect on transistor performance. Due to process variations, the properties of transistors can change to the extent that the functionality of the circuit is compromised. Accounting for the process variations and designing circuits that can guarantee functionality under these variations is crucial.

Scaling down the supply voltage is an effective way of reducing the energy consumption. At the lowest supply voltages, the transistors start to operate in the weak inversion region. Here, the current depends exponentially on the applied voltage and the transistor parameters. This exponential relationship makes that low voltage operation extremely sensitive to variations. The result is a highly unpredictable circuit and system performance which is hard to rely on in any kind of application. While accounting for the worst case variations beforehand is *a* possible solution, it undermines the efforts made to improve energy consumption and operating speed. Improving the predictability of the microcontroller under process variations and potentially operating "better than the worst case" is another main challenge here.

The collection of variations present in advanced CMOS technology, combined with the complexity of the modern microcontroller system, makes the design process a lot more difficult. A lean design flow that enables the designer to manage this complexity is mandatory. Using commercial electronic design automation (EDA) tools with a standard cell implementation flow results in an efficient design process that is industry-applicable and can facilitate today's time-to-market requirements. Combining such an efficient design flow with all the problems variation-prone ultra-low voltage operation brings to the table is another main challenge.

This work tries to combine the described challenges that energy-efficient microcontrollers face in sub-micron CMOS technologies. An ultra-low energy consumption with fast enough performance while being variation-resilient is the triple combination this work targets. Ideally, this system is realized using an efficient design process that helps the designer to improve the system as much as possible. To accomplish this, variation-resilient building blocks and design techniques to operate at ultra-low voltage are presented. This results in the efficient implementation of several microcontroller prototypes fabricated in 40 nm CMOS technology that achieve state-of-the-art performance and ultra-low energy consumption.

This chapter gives a short overview of the main challenges faced in this work, and thereby introduces the remaining Chaps. 2–7. Section 1.2 introduces the concept of minimum energy operation. It is the main target for everything developed in this work. Section 1.3 discusses CMOS technology and the problems it faces when operated for minimum energy. Section 1.4 introduces efficient design through a standard cell based very-large-scale-integration (VLSI) design flow. Variations and their influence is the topic of Sect. 1.5. Section 1.6 discusses current and possible future applications for microcontrollers and other systems that benefit from minimum energy operation and variation-resilient design. These sections set out the main challenges this work faces. Section 1.7 derives the goals of this work. Finally, Sect. 1.8 sets the outline of this work.

1.2 Energy-Efficient Operation

Traditionally, design optimization techniques have targeted to achieve the fastest possible operation. For a given CMOS technology, this typically means operating it at the nominal supply voltage of that technology. Minimum delay operation facilitated by nominal supply voltage operation is therefore said to target the minimum delay point (MDP), shown in Fig. 1.1. CMOS technology scaling continues to shrink transistors to smaller minimum feature sizes, thereby enabling faster operation or more complex functionality. The influence on power consumption is substantial. As the formula for the total power consumption (P_{tot}) of a digital system describes (see Eq. 1.1 [19]), both the static (P_{static}) and the dynamic power consumption (P_{dynamic}) are proportional to the supply voltage (V_{dd}). Static power consumption also depends on the leakage current (I_{leak}). The dynamic power consumption is, apart from the activity (α) and capacitance (C), also proportional to the clock frequency (f_{clk}). The combination of these two effects makes that a high (nominal) V_{dd} and a high clock frequency inevitably results in a high power consumption.

$$P_{\text{total}} = \underbrace{V_{\text{dd}} \cdot I_{\text{leak}}}_{P_{\text{static}}} + \underbrace{\alpha \cdot f_{\text{clk}} \cdot C \cdot V_{\text{dd}}^2}_{P_{\text{dynamic}}} \tag{1.1}$$

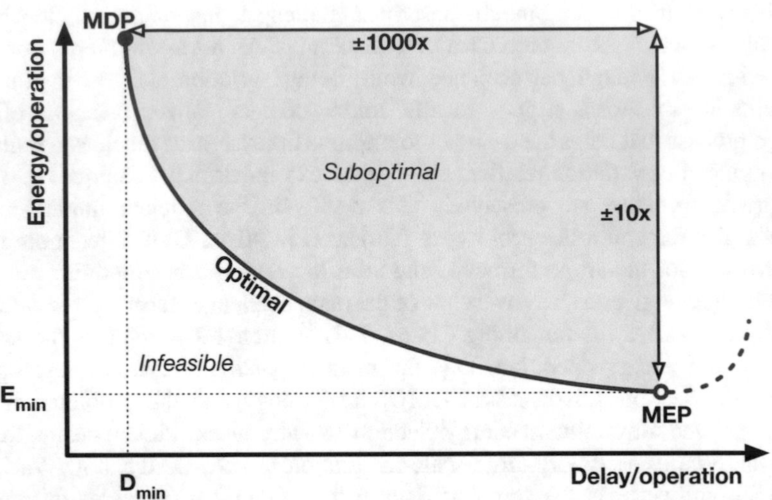

Fig. 1.1 Energy delay trade-off in design optimization, pictured as in [16]. Traditionally the high-energy low-delay region (MDP) was targeted. More recently, the high-delay low-energy region (MEP) is of interest as well

Obviously, a high clock frequency also makes for a high throughput, since every clock period delivers computational results. Moreover, as transistors shrink, their capacitance and leakage current decreases which decreases power consumption. Because of this nuance, it helps to look at the total energy consumption (E_{total}, see Eq. 1.2 [19]). Dynamic energy (E_{dynamic}) depends quadratically on V_{dd}, which means it benefits significantly from lower V_{dd} operation. Static energy (E_{static}) depends on the V_{dd}, I_{leak} and the clock period (T_{clk}), which both also depend on V_{dd}. Static energy typically increases with lower V_{dd} operation because the leakage current is integrated over a longer clock period [33]. The static and dynamic component of the energy consumption result in a trade-off: when decreasing V_{dd}, the dynamic energy drastically reduces, but is counteracted by the increase in static energy. The total energy reaches an optimum: the minimum energy point (MEP) [33]. Operation at a higher V_{dd} than $V_{\text{dd,MEP}}$ increases the total energy due to higher dynamic energy, while operation at a lower V_{dd} increases the total energy due to higher static energy.

$$E_{\text{total}} = \frac{P_{\text{total}}}{f_{\text{clk}}} = \underbrace{V_{\text{dd}} \cdot I_{\text{leak}} \cdot T_{\text{clk,max}}}_{E_{\text{static}}} + \underbrace{\alpha \cdot C \cdot V_{\text{dd}}^2}_{E_{\text{dynamic}}} \tag{1.2}$$

Designing for this MEP produces significantly different results than designing for the MDP. Figure 1.1 shows how the MDP and MEP relate. The MDP minimizes delay per operation at a high energy cost, while the MEP minimizes energy per operation at a high delay penalty. In between is a *pareto* front that traces the best

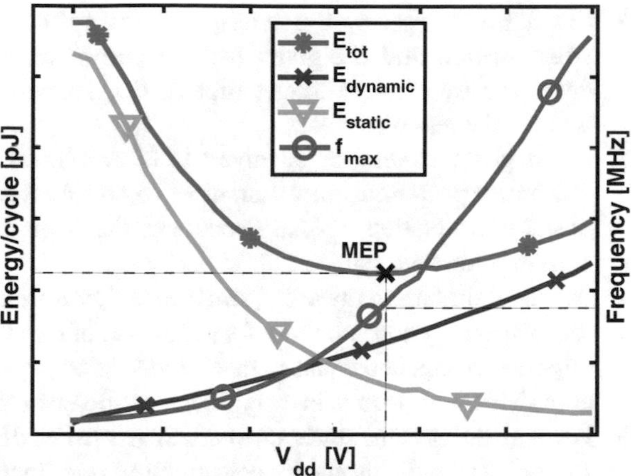

Fig. 1.2 Energy breakdown and speed performance of one of the microcontroller prototypes presented in Chap. 4

possible energy consumption for the desired speed performance, or the best possible speed performance for a given energy budget. Operation at a lower V_{dd} than $V_{dd,MEP}$ always results in a higher energy per operation. From an energy perspective, such operation is inefficient, since more energy is consumed for the same operation at a lower speed. From a power perspective, this could make sense since both static and dynamic power decrease, resulting in a lower total power consumption. Other metrics such as the energy-delay product (EDP) are frequently used to demonstrate the trade-off between energy and speed performance represented by this pareto front.

This work develops several microcontroller prototypes with the aim to operate at or in close proximity to the minimum energy point. The $V_{dd,MEP}$ typically results in operation at or around the threshold voltage V_T. For this reason, circuits that operate in this region are often called near-threshold circuits [6, 16, 24, 31]. As will be shown throughout this work, simply decreasing the supply voltage without consideration for its other effects will not result in optimal energy operation. The entirety of the design process, from technology choice and transistor behaviour, all the way up to architectural considerations and program code, influences the energy consumption, speed performance and supply voltage of the minimum energy point. Figure 1.2 looks at the energy breakdown of one of the prototypes developed in the later chapters of this work. Looking at such a breakdown, some key considerations valuable to this work regarding the energy consumption of a digital system are given below.

- For a given system and a given task, **the minimum energy point is uniquely defined by its supply voltage, operating speed and energy consumption.** All three should be taken into account when considering minimum energy operation.
- The minimum energy point is **the most efficient point for a fixed system and a fixed task.** There is no operating point that can perform the same group of operations at any other V_{dd} with a lower energy budget.

- For a given system, **the energy consumption considers both the power consumption and the given task to complete**. Power consumption is easily decreased by taking a longer time to do that task, while energy consumption will not decrease.
- For a given system and a given task, **minimum energy consumption and minimum power consumption are two completely different things.** Minimum power consumption typically occurs at the lowest V_{dd} while minimum energy consumption does not.
- At the minimum energy point, **static and dynamic energy contribute equally** to the total energy consumption. As a function of the supply voltage, their curves are opposite in sign and equal in magnitude. In any system with any task, operation at the MEP thus results in a significant static energy consumption.
- For a given system, **static power consumption does not increase with lower V_{dd} operation.** Static energy consumption does increase, as a result of the longer clock period over which the leakage current is integrated.
- **Speed performance is proportional to** V_{dd}, which means a higher $V_{dd,MEP}$ enables faster minimum energy operation. A way to achieve a significantly higher speed MEP for the same system is by changing technological aspects like the threshold voltage.
- **All the variables that influence static and dynamic energy contribute to the minimum energy point.** Apart from the supply voltage, also the leakage current, clock period, activity and load capacitance of the system impact the minimum energy point.
- For a given operating frequency and system, **the best operating point in terms of energy is the lowest V_{dd} that allows operation at that speed**. It is essential to design for a MEP that achieves the target operating frequency.
- **Operation at the MEP is not enough.** Any system functional at a low enough V_{dd} demonstrates a minimum energy point. The best system realizes the lowest minimum energy point for the given task at exactly the right target frequency.

1.3 CMOS Technology

The CMOS technology used to implement a minimum energy digital system is of major importance. Since the MEP is typically located at a significantly lower supply voltage than the nominal V_{dd}, designing for minimum energy is a challenge. Transistor functionality differs significantly from nominal V_{dd} operation. Near-threshold operation typically does not result in a fully inverted transistor channel, which decreases drive current (I_{on}) substantially. As a result, a lower speed performance can be expected. Moreover, the leakage current (I_{off}) does not decrease proportionally. As a result, the I_{on}/I_{off} current ratio becomes significantly smaller. For the CMOS technologies employed in this work, another important property is

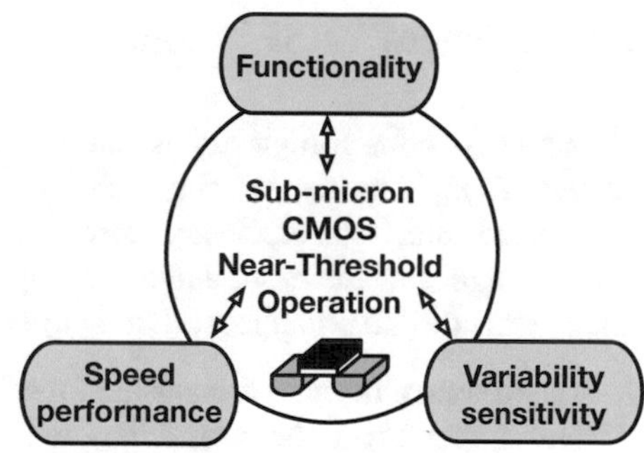

Fig. 1.3 Near-threshold operation in sub-micron CMOS technology results in functional problems, speed degradation and variability sensitivity

the relationship between the complementary pMOS-nMOS pair. When applied in near-threshold, the pMOS drive current degrades significantly more than the nMOS current. This means a logic gate designed for nominal supply voltage operation will not function properly at near-threshold supply voltages (Fig. 1.3).

As CMOS technologies shrink to smaller feature sizes, most transistor properties that influence digital systems improve. Both the area and the gate capacitance decrease. But as transistors approach nanometer sizing, their properties also become highly variable. Systematic variations on all devices of a single chip or wafer can occur due to unintended but inevitable fluctuations in the production process. These kind of variations are typically referred to as *inter-die variations* or process corners. On top, random imperfections in the device manufacturing can occur as well. They result in different transistor parameters, even when they were designed to be identical. As they influence each transistor independently, they are typically referred to as *intra-die variations* or mismatch. Operation at the near-threshold voltage makes transistors highly susceptible to both these types of variations. The already lower I_{on} and I_{off} both have a higher relative spread, since the change in parameters has a higher impact. These variations are reflected in the logic gates and the systems they compose, making operating speed and energy consumption variable quantities.

When building gates that perform at near-threshold voltages, these effects have to be taken into account. Key challenges are guarding the logic functionality, overcoming the reduction in operating speed and overcoming the high susceptibility to process variations. In this work, the necessary building blocks to enable correct functionality at near-threshold voltages are examined. Simulations show how the speed of logic gates degrades as a result of near-threshold operation. The effects of process variations are equally accounted for. For logic functions, we rely for the bigger part on the transmission gate building block as presented in [24]. Its good functional, speed and variation-resilient properties are demonstrated through transistor and gate level simulations and system level measurements.

1.4 Efficient VLSI Design

Managing system complexity is one of the biggest challenges in modern VLSI design. Today's system-on-chips (SoC) combine multiple processors, memories, high-speed interfaces, dedicated accelerators, etc. into one big system. A modern VLSI design flow therefore defines multiple (typically five) levels of abstraction to manage the design complexity and facilitate a fast design cycle.

1. **Architecture design:** designs the functions of the system. For a microcontroller these typically are the instruction set, the register set, the memory model, etc.
2. **Micro-architecture design:** defines how the architecture is partitioned into functional units. For a microcontroller this means different flavours that change up gate count, benchmark performance and power trade-offs such as single or multi-cycle multiplication, etc.
3. **Logic design:** defines the construction of the functional units. A function like an adder can be implemented in different ways: ripple carry, carry look-ahead, carry select, etc.
4. **Circuit design:** defines how transistors are used to implement the logic functions. Typical options are conventional static CMOS, transmission gate, domino or pass gate logic.
5. **Physical design:** defines the layout of the system, with all of its physical implementation considerations such as cell placement, routing and clock tree synthesis.

Minimum energy design is influenced by every abstraction level. The best possible system can only be achieved when every design choice serves the minimum energy paradigm. Moreover, all choices at every level almost always influence all other levels. This makes the job of the designer more challenging than ever.

In most cases, circuit design and physical design are facilitated using a standard cell design flow. A standard cell library consists of a predefined collection of cells with different transistor implementations, logic functions and drive strengths. Their timing and power behaviour are pre-characterized according to a range of input signal slew rates and output signal load capacitors. This provides enough information to predict their behaviour when they are combined into a digital system. Logic synthesis, static timing analysis, power analysis and a place and route based layout are possible through standard cell design. This facilitates large scale implementation, design behaviour predictability and a fast design cycle.

Transmission gate building blocks as described in [24] interfere with levels 4 and 5 of the typical standard cell based design flow. They do not readily allow standard cell based implementation with commercial electronic design automation (EDA) software. As a result, the proposed approach lacks much of the good properties of the typical design flow and its EDA tools bring: logic synthesis, static timing analysis, power analysis, place and route and fast and efficient design.

This work enables standard cell based VLSI design with variation-resilient and near-threshold functional transmission gate building blocks. The generation and characterization of a transmission gate based standard cell library is discussed. A commercial tool flow based design strategy augmented with a few necessary interventions enables all the benefits of standard cell based VLSI design. These interventions are facilitated using automated scripts that do not compromise the design time or complexity. Additionally, timing and power analysis at multiple supply voltages can provide enough insight to make the necessary minimum energy considerations. Moreover, the most recent innovations in the typical VLSI design flow and its tools are focused on power- and variability-aware implementation. Leveraging these to the best of their abilities and possibly more is a perfect match for the research presented in this work.

1.5 Tackling Variations

Advanced sub-micron CMOS technology demonstrates a high variability. Both inter-die and intra-die process variations skew transistor parameters and thus their performance. For the designer, process variations are there as given. Transistor parameters will vary under any circumstance and variation will increase when transistor sizes shrink. In circuit design, designers rely on the models provided with the technology that allow simulation of mismatch and corner conditions. At the circuit level, these models can be used to verify the variation resiliency of the circuit performance. On top of process variations, conditions like ambient temperature, circuit ageing or voltage droop affect circuit behaviour as well. In guaranteeing the performance of a digital system, both process variations and other effects should be taken into account. Creating gate level building blocks that are as variation-resilient as possible is the first challenge here.

The modern VLSI design flow incorporates the process variation models. Typically, a multitude of standard cell library conditions is provided, each representing a single process, voltage and temperature condition. Such a library is accompanied by the characterized timing and power behaviour of each standard cell under those conditions. Because each of these parameters can vary independently, this results in numerous possible combinations: every process corner can occur at every temperature and every voltage condition. To guarantee circuit functionality and performance under each of these conditions, every possible combination has to be checked, and the circuit is co-optimized for each of them. However, all of these conditions don't always occur: a circuit might be fabricated in good process conditions and could therefore perform a lot better than the circumstances it was designed for. Temperature or voltage conditions are typically overestimated "just to make sure". Designing for all these conditions thus incorporates a lot of overhead (or margin). Moreover, in co-optimization for an increasing number of conditions, the design time increases super-linearly. Managing this design process, accounting for all of the possible conditions, while minimizing overhead, is the main challenge for variation-resilient VLSI design.

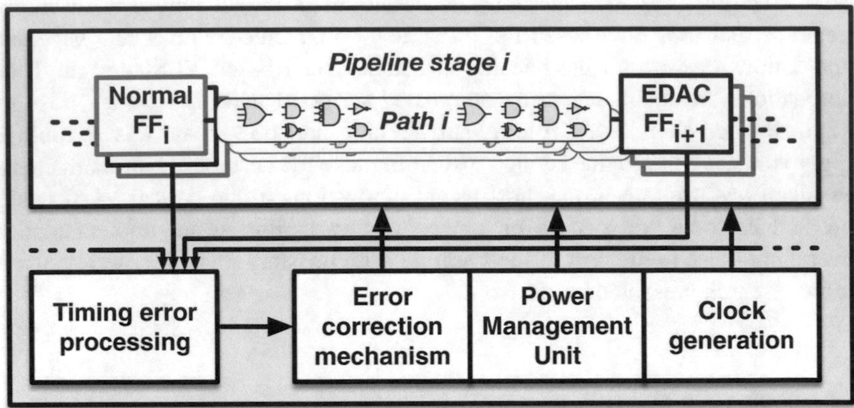

Fig. 1.4 Near-threshold system variations can be overcome by autonomous in-circuit timing monitoring, which facilitates operation near the point-of-first-failure

Because near-threshold operation is even more susceptible to variations, simply guaranteeing performance under the worst case conditions is not enough. Despite the co-optimization capabilities of the VLSI design flow, near-threshold operation results in a huge overhead. When accounting for all conditions, the resulting margins grow to the extent that they neutralize the energy gains near-threshold operation aims for. Since process variations make it so that each individual chip has different properties, most chips will be able to perform much better than the worst case (BTWC). Abandoning the worst case operating point and operating at the edge of where the chip will actually fail is much more preferred. This so-called point-of-first-failure (PoFF) is the ideal scenario and eliminates almost all margins. However, process variations make it so that each chip has a different PoFF, and in the field changes (such as temperature variations) continuously shift the PoFF. Because variation-resilient building blocks only go so far, an architecture that enables per-chip PoFF operation is the next step.

In this work this per-chip PoFF operation is realized with a system-level control loop that compensates for variations and thereby guarantees operation at the PoFF of each individual chip for a certain frequency target. A system overview is shown in Fig. 1.4. Because near-threshold circuits are so susceptible to (intra-die) variations, there is no choice but to monitor circuit performance in continuously in real-time, inside the actual circuit (in situ). For a sequential flip-flop based pipeline as used in the microcontroller, this is done by looking at the data arrival of the logic path in the flip-flop. In margined conditions, this data arrives early. When the system fails, this data arrives late. Just-in-time arrival means operation at the edge of the PoFF. Scaling the data arrival time accordingly can be done by scaling the supply voltage. Gathering the timing arrival information from the flip-flops at the system level enables supply voltage control to this end.

Operation at the edge of the PoFF implies risk, since a (sudden) change in conditions may result in system failure. Therefore, such a system needs a mechanism to correct this failure. Because of this, the combination of a (late) data arrival detection with a correction mechanism is often referred to as an error detection and correction (EDAC) system. While this principle has been demonstrated extensively before in literature [3–5, 7, 10, 12, 34, 35], the implementation trade-offs change considerably for near-threshold operation. The main challenge is to enable near-threshold operation, and guard the overhead such an EDAC system brings along. A system level correction mechanism typically results in a large overhead. In this work, correction is realized through error masking, which forwards the correct data despite its late arrival at the flip-flop, similar to the mechanism of time-borrowing in latch-based pipelines. Moreover, data arrival detection is only done a subset of the total amount of flip-flops to limit the overhead. This is realized through a thorough statistical analysis of the delay of the different logic paths that make up the system. Overall, the key consideration is that the architectural additions for EDAC operation should not result in more overhead than the margins for worst case conditions would add to the overall energy consumption.

1.6 Applications

The widespread use of microcontrollers in embedded devices results in a long list of applications. Microcontrollers are being used in almost every device that incorporates some level of programmable electronics. The beginning of this chapter mentioned only a few. However, efforts to reduce the energy consumption of the microcontroller or other digital systems are not always adopted. For mains-powered applications the digital system energy consumption is often negligible, certainly when they incorporate other systems (e.g. electro-motors) that consume orders of magnitude more energy. Ultra-low energy operation is more suited for applications where the majority of the energy consumption goes to the digital processing, such as digital signal processing (DSP) applications, and the speed reduction is tolerable. For battery-powered applications in this field, the energy budget is usually a lot tighter. Especially when periodical battery refill is impractical, expensive or simply impossible. Systems that rely on energy harvesting for their operation have an even tighter energy budget. In these two categories, applications at hand are wireless sensor nodes, biomedical implants, hearing aids, etc.

To some extent, the techniques to realize minimum energy consumption presented in this work are applicable to any digital circuit. If the speed degradation that occurs when operating in the near-threshold region is tolerable, the energy consumption of almost any digital logic circuit will benefit. The microcontroller prototypes presented further in this work are a representative building block for a variety of digital systems. A major merit of this work is the high speed-low energy combination it realizes. This can enable new applications that were out of scope for near-threshold operation up till now.

The continuous advancements of CMOS technology will inevitably result in higher variations. Although foundries are continuously improving the CMOS production process, smaller devices will still be variation-prone. Variation-resilient design techniques and building blocks are gaining attention, and will continue to do so. The techniques presented in this work add to this field.

In the same context, performance monitoring techniques such as error detection and correction are gaining attention. Error detection and correction with instruction replay has been reported in commercial processors for enterprise servers and high-performance computing applications. Bostian [2] reports the use of error correcting code, parity bits and arithmetic operation residues to detect soft errors (e.g. through an α-particle) and instantly correct them. The high-end requirements of this processor justifies the design effort and implementation overhead of error detection and correction. As requirements for more low-end systems tighten, the threshold to implement these kind of performance monitoring systems lowers. Moreover, rarely do applications tolerate performance degradation in favour of energy consumption. The control loop that error detection systems enable can help near-threshold systems meet a given performance under all conditions.

1.7 Goals of This Work

The previous sections of this chapter introduced the major parts of this work, and formulated the most important challenges. In one sentence, this translates to **variation-resilient digital systems that operate under minimum energy conditions with decent speed performance, implemented using an efficient design flow and in CMOS technology**. More specifically, the following key challenges are stated:

- **Functionality:** Near-threshold operation is the key to minimizing energy. Enabling near-threshold functionality is a non-trivial challenge and is well discussed in Chap. 2.
- **Speed performance:** With near-threshold operation comes speed degradation. Providing a high enough performance without compromising energy consumption is key. Building blocks that enable speed performance are discussed in Chap. 2, while system level adaptations that guarantee a given speed performance are discussed in Chap. 6.
- **Variability-resilience:** Near-threshold operation is highly susceptible to variations. Building blocks that are particularly variation-resilient are discussed in Chap. 2. Chapter 3 focuses on a design flow that guards these good properties. Mitigating process variations on the transistor level only gets you so far. Chapter 6 demonstrates system level adaptations that compensate for process variations and thereby guarantee system performance.
- **Efficient VLSI design:** An EDA tool supported design flow is the only feasible way to design large digital systems. Near-threshold operation is not always

compatible with the typical standard cell flow. Chapter 3 discusses the generation of a standard cell library and the design flow used to implement a digital system with these standard cells. Chapter 6 extends this for error detection and correction. The most recent innovations in VLSI design aim to manage power consumption and design margins. Exactly these topics are the key goals of this work regarding VLSI design.

- **Insights in minimum energy operation:** With efficient design comes control. Timing analysis, power analysis and a fast design iteration cycle provide system level insight in the static and dynamic energy consumption. This improves the predictability of the minimum energy point and enables the designer to target this point. Chapter 4 demonstrates such insights for the microcontroller systems that were implemented there.
- **Point-of-first-failure operation:** Variation-resilient building blocks only go so far. To guarantee a predefined performance with minimal overhead, architecture level adaptations that monitor the systems performance in real time are preferred. Timing error detection and correction can go a long way in enabling this, but also introduces overhead. Moreover, near-threshold error detection and correction poses severe challenges. To the authors best knowledge, no literature enables near-threshold error detection and correction with a flip-flop based pipeline. Chapter 5 discusses this technique at length, Chap. 6 demonstrates its application in a near threshold microcontroller prototype and discusses the energy considerations that were made.

The state-of-the-art comparisons in Chaps 4 and 6 situate the prototypes developed in this work in relation to other recent published work. Taking a closer look at some of these recent publications, the first key observation is that some works do not even achieve the true minimum energy point because functionality at such low supply voltages is compromised [13]. The building blocks used in this work relate for the bigger part to the works published in [20–23]. These works do realize variation-resilient sub- or near-threshold operation combined with a high-speed performance, but at a huge design cost: full custom transistor-based design and SPICE-level simulations. An efficient VLSI design is what lacks these works.

Looking at recent near-threshold microcontroller implementations, a key challenge appears to be speed performance. From [14, 15, 17, 18], only one reports a MHz-range operating speed at the MEP, and achieves this through an advanced 14 nm Tri-Gate CMOS technology. To accurately compare variation-resilience, corner-lot production samples are necessary, combined with enough measured dice to gain insight in the process variation sensitivity that skews the measurements results. None of these works do such a thing, for the same reason that this is not done in this work. Only industry-level fabrication volume with automated testing allows such an analysis. The reporting of multiple measurements of different dice has been well adopted, and gives some insights into variation-resilience. Finally, none of these works implement better-than-worst-case operation of some kind.

When looking at the literature that does apply better-than-worst-case operation through EDAC, near-threshold operation becomes a problem. While the error detection and correction technique typically aims to reduce energy consumption through supply voltage scaling, [3–5, 7, 12, 34, 35] do not function in the near-threshold operating region. Kim and Seok [10] is the only work that does enable near-threshold operation, but relies on a two-phase latch based pipeline. This thesis accomplishes the same in a flip-flop based design.

The prototypes developed in this work improve the state of the art on most of these parts with measured results. They combine variation-resilient building blocks with an efficient design flow. As a result, they achieve a high-speed performance at the minimum energy point without compromising energy consumption. Moreover, the final prototype combines these aspects with a system level control loop that senses in-circuit speed-performance variations in real time and compensates them.

1.8 Outline

Chapter 2 explores the key aspects of near-threshold operation in advanced sub-micron CMOS technology. Building blocks to realize functional and fast variation-resilient logic gates are discussed, as well as architectural considerations that help realize minimum energy consumption.

Chapter 3 presents an efficient VLSI design flow that leverages near-threshold operation to its fullest. By looking at typical standard cell based design with commercial EDA tools, this chapter defines the challenges this work faces. They are overcome by augmenting the design flow with the necessary intermediate steps.

Chapter 4 demonstrates the near-threshold building blocks and efficient design flow by implementing two microcontroller prototypes. Prototype 1 functioned as a proof-of-concept and already achieves state-of-the-art performance. Prototype 2 improves on this and achieves even better state-of-the-art results.

Chapter 5 looks at error detection and correction techniques. A wide range of literature is available. The most important aspects are highlighted and conclusions for near-threshold error detection and correction operation are drawn.

Chapter 6 puts the considerations of Chap. 5 to practice. The microcontroller prototype from Chap. 4 is augmented with an error detection and correction system, making it *timing error-aware*. The architecture, circuits, implementation details and measurement results of this system are presented.

Chapter 7 draws the conclusion of this work and its contribution to state of the art. Suggestions for future improvements are made.

Finally, the information presented in the rest of this work has been published in the following papers: [25–30].

References

1. Ashouei, M., Hulzink, J., Konijnenburg, M., Zhou, J., Duarte, F., Breeschoten, A., Huisken, J., Stuyt, J., de Groot, H., Barat, F., David, J., Van Ginderdeuren, J.: A voltage-scalable biomedical signal processor running ECG using 13 pJ/cycle at 1 MHz and 0.4 V. In: IEEE International Solid-State Circuits Conference Digest of Technical Papers (ISSCC), pp. 332–334. IEEE, New York (2011)
2. Bostian, S.: Rachet up reliability for mission-critical applications: Intel® instruction replay technology. White Paper, 48 (2013)
3. Bowman, K.A., Tschanz, J.W., Kim, N.S., Lee, J.C., Wilkerson, C.B., Lu, S.L.L., Karnik, T., De, V.K.: Energy-efficient and metastability-immune resilient circuits for dynamic variation tolerance. IEEE J. Solid-State Circ. **44**(1), 49–63 (2009)
4. Bull, D., Das, S., Shivashankar, K., Dasika, G.S., Flautner, K., Blaauw, D.: A power-efficient 32 bit ARM processor using timing-error detection and correction for transient-error tolerance and adaptation to PVT variation. IEEE J. Solid-State Circ. **46**(1), 18–31 (2011)
5. Das, S., Tokunaga, C., Pant, S., Ma, W.H., Kalaiselvan, S., Lai, K., Bull, D.M., Blaauw, D.T.: RazorII: in situ error detection and correction for PVT and SER tolerance. IEEE J. Solid-State Circ. **44**(1), 32–48 (2009)
6. Dreslinski, R.G., Wieckowski, M., Blaauw, D., Sylvester, D., Mudge, T.: Near-threshold computing: reclaiming Moore's law through energy efficient integrated circuits. Proc. IEEE **98**(2), 253–266 (2010)
7. Fojtik, M., Fick, D., Kim, Y., Pinckney, N., Harris, D.M., Blaauw, D., Sylvester, D.: Bubble Razor: eliminating timing margins in an ARM cortex-M3 processor in 45 nm CMOS using architecturally independent error detection and correction. IEEE J. Solid-State Circuits **48**(1), 66–81 (2013)
8. IC-Insights: MCU Market Forecast (2016). http://www.icinsights.com/news/bulletins/MCU-Market-Forecast-To-Reach-Record-High-Revenues-Through-2020/
9. Jayakumar, H., Lee, K., Lee, W.S., Raha, A., Kim, Y., Raghunathan, V.: Powering the internet of things. In: Proceedings of the International Symposium on Low Power Electronics and Design (ISLPED), pp. 375–380. ACM Press, New York (2014)
10. Kim, S., Seok, M.: Variation-tolerant, ultra-low-voltage microprocessor with a low-overhead, within-a-cycle in-situ timing-error detection and correction technique. IEEE J. Solid-State Circ. **50**(6), 1478–1490 (2015)
11. Kim, H., Kim, S., Van Helleputte, N., Artes, A., Konijnenburg, M., Huisken, J., Van Hoof, C., Yazicioglu, R.F.: A configurable and low-power mixed signal SoC for portable ECG monitoring applications. IEEE Trans. Biomed. Circ. Syst. **8**(2), 257–267 (2014)
12. Kwon, I., Kim, S., Fick, D., Kim, M., Chen, Y.P., Sylvester, D.: Razor-Lite: a light-weight register for error detection by observing virtual supply rails. IEEE J. Solid-State Circ. **49**(9), 2054–2066 (2014)
13. Lallement, G., Abouzeid, F., Cochet, M., Daveau, J.M., Roche, P., Autran, J.L.: A 2.7 pJ/cycle 16 MHz, 0.7 µW Deep Sleep Power ARM Cortex-M0+ Core SoC in 28 nm FD-SOI. IEEE J. Solid-State Circ. **53**, 1–13 (2018)
14. Lim, W., Lee, I., Sylvester, D., Blaauw, D.: Batteryless sub-nW Cortex-M0+ processor with dynamic leakage-suppression logic. In: IEEE International Solid-State Circuits Conference Digest of Technical Papers (ISSCC), pp. 1–3. IEEE, New York (2015)
15. Luetkemeier, S., Jungeblut, T., Porrmann, M., Rueckert, U.: A 200 mV 32b subthreshold processor with adaptive supply voltage control. In: IEEE International Solid-State Circuits Conference Digest of Technical Papers (ISSCC), pp. 484–486. IEEE, New York (2012)
16. Markovic, D., Wang, C., Alarcon, L., Tsung-Te Liu, Rabaey, J.: Ultralow-power design in near-threshold region. Proc. IEEE **98**(2), 237–252 (2010)
17. Myers, J., Savanth, A., Gaddh, R., Howard, D., Prabhat, P., Flynn, D.: A subthreshold ARM Cortex-M0+ Subsystem in 65 nm CMOS for WSN applications with 14 power domains, 10T SRAM, and integrated voltage regulator. IEEE J. Solid-State Circ. **51**(1), 31–44 (2016)

18. Paul, S., Honkote, V., Kim, R.G., Majumder, T., Aseron, P.A., Grossnickle, V., Sankman, R., Mallik, D., Wang, T., Vangal, S., Tschanz, J.W., De, V.: A sub-cm3 energy-harvesting stacked wireless sensor node featuring a near-threshold voltage IA-32 microcontroller in 14-nm tri-gate CMOS for always-ON always-sensing applications. IEEE J. Solid-State Circ. **52**(4), 961–971 (2017)
19. Rabaey, J., Chandrakasan, A.P., Nikolic, B.: Digital Integrated Circuits: A Design Perspective, 2nd edn. Pearson Education Inc., London (2003)
20. Reynders, N., Dehaene, W.: A 190 mV supply, 10 MHz, 90 nm CMOS, pipelined sub-threshold adder using variation-resilient circuit techniques. In: IEEE Asian Solid-State Circuits Conference (A-SSCC), pp. 113–116. IEEE, New York (2011)
21. Reynders, N., Dehaene, W.: Variation-resilient building blocks for ultra-low-energy sub-threshold design. IEEE Trans. Circ. Syst. II Express Briefs **59**(12), 898–902 (2012)
22. Reynders, N., Dehaene, W.: Variation-resilient sub-threshold circuit solutions for ultra-low-power Digital Signal Processors with 10 MHz clock frequency. In: 38th IEEE European Solid-State Circuits Conference (ESSCIRC), pp. 474–477. IEEE, New York (2012)
23. Reynders, N., Dehaene, W.: A 210 mV 5 MHz variation-resilient near-threshold JPEG encoder in 40 nm CMOS. In: IEEE International Solid-State Circuits Conference Digest of Technical Papers (ISSCC), pp. 456–457. IEEE, New York (2014)
24. Reynders, N., Dehaene, W.: Ultra-Low-Voltage Design of Energy-Efficient Digital Circuits (Springer, Leuven, 2015)
25. Reyserhove, H., Dehaene, W.: A 16.07 pJ/cycle 31MHz fully differential transmission gate logic ARM Cortex M0 core in 40 nm CMOS. In: 42nd IEEE European Solid-State Circuits Conference (ESSCIRC), pp. 257–260. IEEE, New York (2016)
26. Reyserhove, H., Dehaene, W.: A differential transmission gate design flow for minimum energy sub-10-pJ/cycle ARM cortex-M0 MCUs. IEEE J. Solid-State Circ. **52**(7), 1904–1914 (2017)
27. Reyserhove, H., Dehaene, W.: Design margin elimination in a near-threshold timing error masking-aware 32-bit ARM Cortex M0 in 40 nm CMOS. In: 43rd IEEE European Solid-State Circuits Conference (ESSCIRC), pp. 155–158. IEEE, New York (2017)
28. Reyserhove, H., Dehaene, W.: Design margin elimination through robust timing error detection at ultra-low voltage. In: IEEE SOI-3D-Subthreshold Microelectronics Technology Unified Conference (S3S), pp. 1–3. IEEE, New York (2017)
29. Reyserhove, H., Dehaene, W.: Margin elimination through timing error detection in a near-threshold enabled 32-bit microcontroller in 40-nm CMOS. IEEE J. Solid-State Circ. **53**, 2101–2113 (2018)
30. Reyserhove, H., Reynders, N., Dehaene, W.: Ultra-low voltage datapath blocks in 28nm UTBB FD-SOI. In: IEEE Asian Solid-State Circuits Conference (A-SSCC), pp. 49–52. IEEE, New York (2014)
31. Wang, A., Calhoun, B.H., Chandrakasan, A.P.: Sub-Threshold Design for Ultra Low Power Systems. Springer, New York (2006)
32. Warneke, B., Pister, K.: An ultra-low energy microcontroller for Smart Dust wireless sensor networks. In: IEEE International Solid-State Circuits Conference Digest of Technical Papers (ISSCC), pp. 316–317. IEEE, New York (2004)
33. Weste, N., Harris, D.: CMOS VLSI Design: A Circuits and Systems Perspective, 4th edn. Addison-Wesley Publishing, Boston (2010)
34. Whatmough, P.N., Das, S., Bull, D.M.: A low-power 1-GHz razor FIR accelerator with time-borrow tracking pipeline and approximate error correction in 65-nm CMOS. IEEE J. Solid-State Circ. **49**(1), 84–94 (2014)
35. Zhang, Y., Khayatzadeh, M., Yang, K., Saligane, M., Pinckney, N., Alioto, M., Blaauw, D., Sylvester, D.: iRazor: current-based error detection and correction scheme for PVT variation in 40-nm ARM Cortex-R4 processor. IEEE J. Solid-State Circ. **53**(2), 619–631 (2018)

Chapter 2
Near-Threshold Operation: Technology, Building Blocks and Architecture

Abstract This chapter explores the key aspects of near-threshold operation. Technology constraints, building blocks and architectural aspects for operating circuits at ultra-low voltage are discussed. All the simulations and prototypes developed in Chap. 2–6 were acquired using a 40-nm CMOS technology. The transistor behaviour of said technology is the base of all the design considerations made further on. Section 2.1 lays out the ground work for this: transistor operating regions, device sizing, FO4 inverter performance and others are presented. These analyses enable an intuitive but surprisingly accurate insight in the microcontroller prototypes developed further in this work.

The logic gate topology used across all the prototypes presented in this work is the differential transmission gate. Section 2.2 looks at the different aspects influencing this choice and compares with other approaches. The VLSI design methodology motivates us to use sequential clock edge triggered pipelines. The flip-flop building block used in this work is briefly discussed in Sect. 2.3, together with some considerations on how it impacts the microcontroller prototypes.

Architectural properties of a digital system equally influence the system's ultra-low voltage performance and minimum energy design target. Pipeline depth and circuit activity are important considerations. Their influence is discussed in Sect. 2.4. The impact of recent advancements in CMOS technology is briefly discussed in Sect. 2.5.

Every section of this chapter briefly sketches the application of the discussed considerations by looking forward to the prototypes of Chaps. 4 and 6. In doing so, the considerations made in this chapter and the conclusions presented in Sect. 2.6 become more tangible.

2.1 Technology

2.1.1 Weak Inversion

Weak-inversion operation, often also referred to as sub-threshold operation, refers to the channel of a MOSFET device not being fully inverted. Figure 2.1 shows the cross section of the typical nMOS and pMOS complementary devices in a

H. Reyserhove, W. Dehaene, *Efficient Design of Variation-Resilient Ultra-Low Energy Digital Processors*, https://doi.org/10.1007/978-3-030-12485-4_2

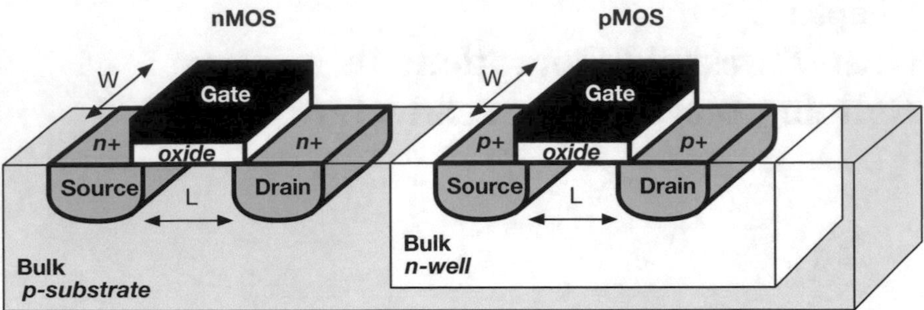

Fig. 2.1 Device cross section of an nMOS and a pMOS transistor in a typical CMOS technology

standard CMOS technology. When increasing the gate–source voltage (V_{gs}) of an nMOS device, assuming the applied drain–source voltage (V_{ds}) is the nominal supply voltage ($V_{dd,nom}$) of the technology, three distinct operating regions can be distinguished in relation to the drain–source current (I_{ds}). **I_{ds} exhibits, respectively, an exponential, quadratic and linear relationship to V_{gs}.** These three regions are demonstrated when looking at the normalized drain–source current of a minimum sized nMOS device of the 40 nm technology of this work, shown in Fig. 2.2. More often, they are referred to as the weak inversion, strong inversion and velocity saturation region. Detailed equations of the relationship between I_{ds} and device voltages, sizing and technology parameters can be found in [23]. A thorough background on MOS device operation is available in [30].

What distinguishes the weak inversion from the strong-inversion region is the exponential relationship between I_{ds} and V_{gs}. The value of V_{gs} at this turnover point is called the **threshold voltage V_T**. The threshold voltage is influenced by various parameters and accounts for a number of sub-micron CMOS technology effects. In [30], V_T is defined as in Eq. (2.1). In practice, V_T is often defined as the gate–source voltage necessary to inflict a predefined drain–source current

$$V_T = \underbrace{V_{T0}}_{1} + \underbrace{\gamma \cdot (\sqrt{\phi_0 + V_{sb}} - \sqrt{\phi_0})}_{2} - \underbrace{\eta \cdot V_{ds}}_{3} - \underbrace{\Delta V_T}_{4} \tag{2.1}$$

1. **V_{T0}**: The zero bias threshold voltage (V_{sb}= 0) [23, 27, 30, 31].
2. **Body Effect:** Modulation of the threshold voltage due to the bulk source voltage V_{sb}, increasing the threshold voltage with increasingly positive V_{sb} [18].
3. **Drain-Induced Barrier Lowering (DIBL):** Accounts for the threshold voltage change in short channel length devices. The source and drain depletion regions widen with increasing V_{ds}. In short channel devices they extend relatively far, thereby influencing the channel and decreasing the threshold voltage [23, 27].
4. **Short Channel Effect:** The proximity of the source and drain depletion regions to the channel influences the threshold voltage to a large extent in short channel devices. As the length decreases, ΔV_T increases and V_T decreases.

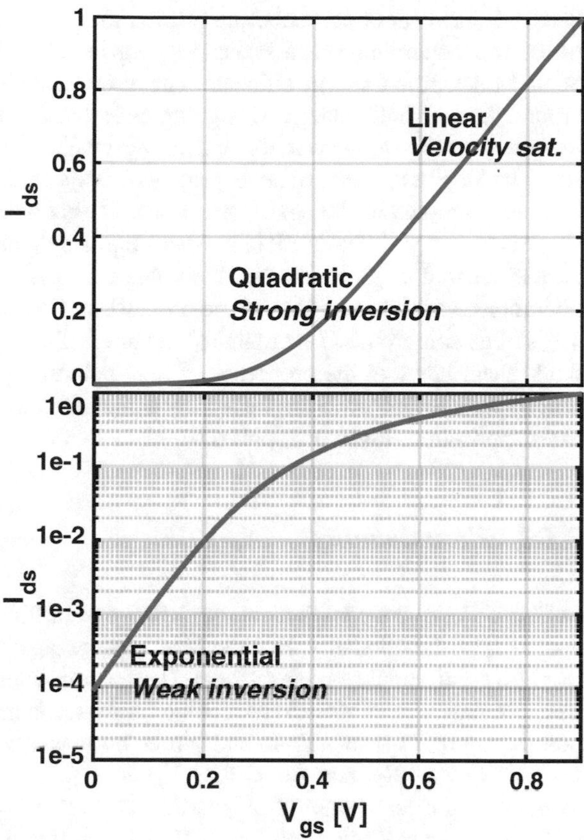

Fig. 2.2 Normalized drain–source current I_{ds} for increasing V_{gs} for a typical low-V_T nMOS device in 40 nm CMOS technology. The linear scale on top shows the quadratic and linear relationship. The log scale on the bottom shows the exponential relationship

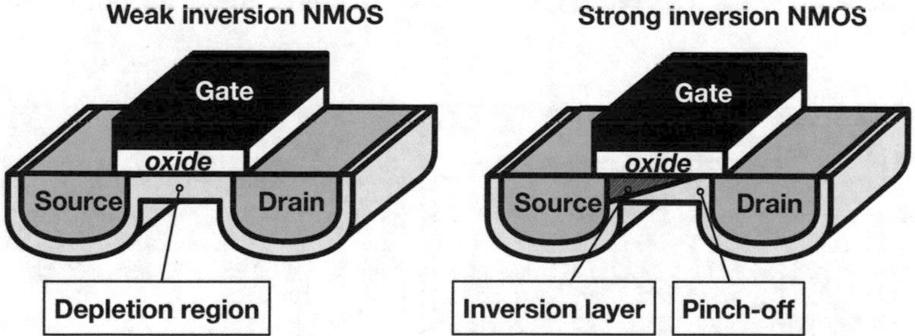

Fig. 2.3 Inversion operation of an nMOS MOSFET. Weak inversion results in a depletion region where diffusion steers current flow. Strong inversion results in a (pinched off) channel where drift controls current flow. Near-threshold operation balances these two

Near-threshold operation is being referred to more recently to describe the operating region at or around the threshold voltage. The channel is not fully inverted and drain–source current is steered by both diffusion and drift (see Fig. 2.3) [30]. As mentioned in Section 2.6.4 of [30], there is little use in defining an exact region here. It is however of particular interest to this work because it offers a better trade-off between operating speed and energy consumption [4, 13], as shown later on in Sect. 2.1.8. Optimal energy efficiency can often be achieved in a relative broad range around the threshold voltage. Designing near-threshold circuits therefore combines properties of the weak-inversion and strong-inversion region. This becomes clearer when looking at the performance, propagation speed and leakage power of the FO4 inverter discussed in the next subsections. In this work, the [0.2 V–0.5 V] voltage range is considered. The MEP of most systems implemented in advanced nanometer CMOS technology is located in this voltage range. The analyses show that 0.2 V operation clearly shares much of the properties of weak-inversion operation. Lower V_{dd}'s often compromise functionality and are difficult to achieve. Higher V_{dd}'s than 0.5 V share much of the properties of nominal supply voltage operation. A MHz-range speed performance can be readily achieved in this voltage range, sufficing for most applications this work applies to.

2.1.2 *Functionality*

In digital logic, the complementary nMOS and pMOS transistors are combined to enable logic functionality. If logic levels are chosen to coincide with the ground (logic *0*) and supply voltage (logic *1*), the transistors reside in one of the two possible states: *on* or *off*. *On* means the transistor is able to conduct the maximum possible current between drain and source by means of a (partly) inverted channel: for nMOS V_{gs}=V_{dd} and for pMOS $V_{sg} = V_{dd}$. *Off* means the transistor is able to conduct the least amount of possible current between drain and source since there is no channel inversion: for nMOS $V_{gs} = 0$ and for pMOS $V_{sg} = 0$. In both cases we assume the $V_{sb} = 0$. Figure 2.4 shows the current for both cases and

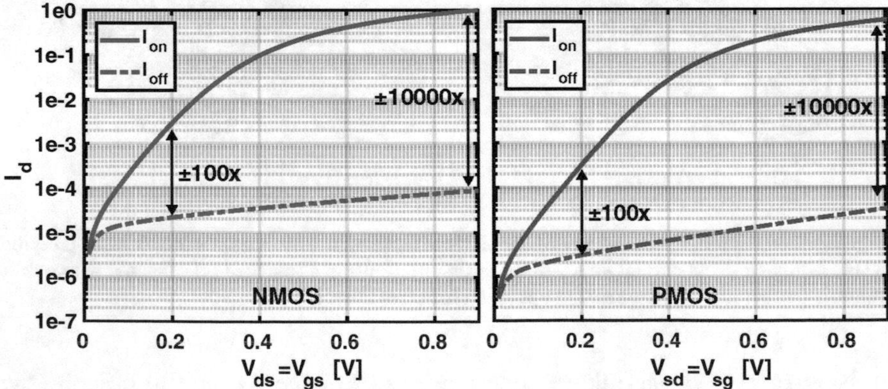

Fig. 2.4 *On* and *off* current for increasing V_{ds} for a minimal low-V_T nMOS (left) and pMOS (right) in 40 nm CMOS technology, normalized to the nominal supply $I_{on,n}$. The I_{on} / I_{off} ratio decreases significantly at lower V_{dd}

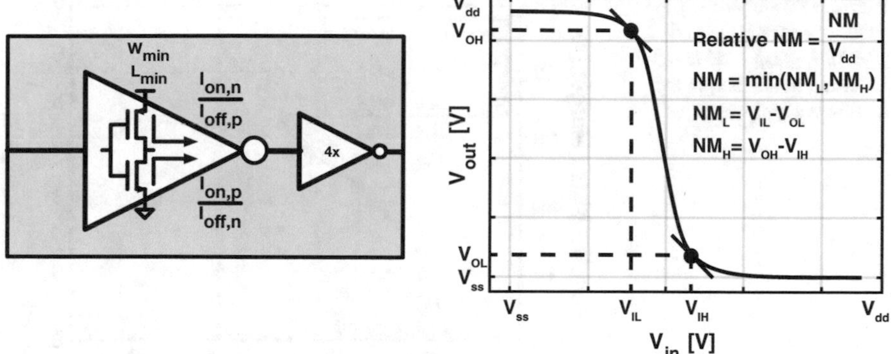

Fig. 2.5 FO4 inverter: logic functionality is realized by balancing drive and leakage current (I_{on}/I_{off}), which is reflected in the noise margin NM

transistors, normalized to the nominal supply $I_{on,n}$. For both transistors, the I_{on} / I_{off} ratio decreases significantly at lower V_{dd}: from 11,200 and 17,700 at 0.9 V to 133 and 108 at 0.2 V. Moreover, at 0.2 V the nMOS I_{on} is far larger than the pMOS I_{on} (8.9× compared to 1.7× at 0.9 V). The same goes for the nMOS and pMOS I_{off} (7.3× compared to 2.7× at 0.9 V). With these ratios, **the transistors can no longer be considered as being *on* or *off*. They simply display a higher or lower I_{ds}.**

The reduced I_{on}/I_{off} ratio means logic functionality is no longer guaranteed when simply combining complementary devices. Logic levels are determined by balancing the current of the pull-up and pull-down device. Since the nMOS I_{off} is comparable to the pMOS I_{on}, the pMOS transistor can no longer drive a logic 1. This breaks the complementary function. In contrast to operation in the strong-inversion region, near-threshold logic has to be ratioed to be functional [13]. This becomes awfully clear when looking at a simple FO4 inverter, shown in Fig. 2.5. Logic functionality is realized by balancing drive and leakage currents, respectively, I_{on} and I_{off} of the pMOS and nMOS transistor. Figure 2.6 shows the noise margin (NM) for the inverter, simulated at different supply voltages. At close to nominal V_{dd}, the I_{on}/I_{off} ratio is high. At lower V_{dd}, the I_{on}/I_{off} ratio deteriorates severely. The noise margin deteriorates at low supplies, eventually breaking functionality. The cure is in increasing the width of the pMOS transistor, thereby balancing the I_{on}/I_{off} ratios and enabling closer to 50% noise margin relative to V_{dd} (RNM). Figure 2.7 shows the best possible relative pMOS sizing and its corresponding relative noise margin across the entire supply voltage range. Because the I_{on} and I_{off} of each device can't be chosen independently, relative noise margin is limited to 35% at the lowest V_{dd}'s. To enable 0.2 V operation, the pMOS requires a relative size of close to 11. Note that this consideration may change when considering different CMOS technologies (see Sect. 2.5).

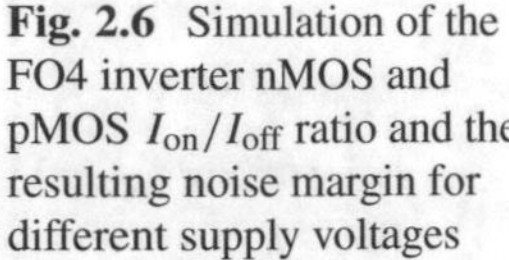

Fig. 2.6 Simulation of the FO4 inverter nMOS and pMOS I_{on}/I_{off} ratio and the resulting noise margin for different supply voltages

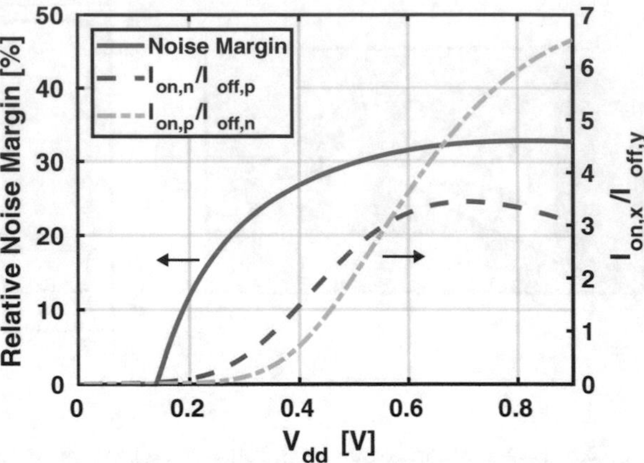

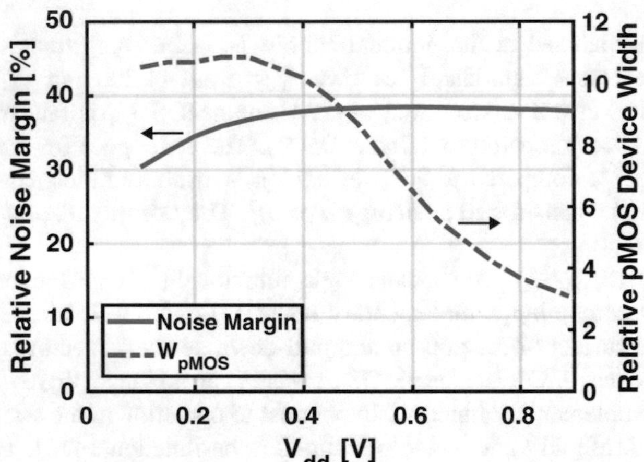

Fig. 2.7 FO4 inverter: larger relative pMOS sizing balances I_{on}/I_{off} ratios and restores the relative noise margin, thus functionality

Key in enabling near-threshold functional operation is thus **balancing the pull-up and pull-down devices**. A good performance metric for operability is noise margin. It indicates how well balanced the on- and off-currents are. The FO4 inverter analysis indicated that balancing **can come at a significant area cost**. This should always be considered when enabling ultra-low voltage operation. Moreover, although the lower drive current may not compromise functionality when pMOS and nMOS are well balanced, the lower drive current does result in a longer propagation delay.

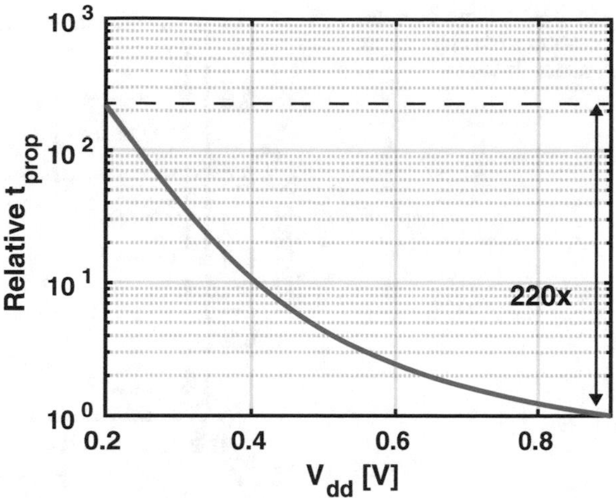

Fig. 2.8 Propagation delay of a FO4 inverter as function of V_{dd}. The delay degrades heavily when reducing V_{dd}

2.1.3 Speed

Near-threshold functionality of complementary logic functions is governed by well-balanced pull-up and pull-down networks, directly determined by the ratio of *on*- and *off*-current. When considering these ratios, one might forget the absolute drive current. As Fig. 2.4 already showed, I_{on} degrades with a factor 10–1000× when comparing nominal V_{dd} operation to 0.2 V operation. The relative propagation delay (t_{prop}) of said FO4 inverter equally reflects this (see Fig. 2.8). In line with the transistor drive current reduction, the propagation delay degrades by two orders of magnitude. Note that the delay degradation becomes much worse below 0.4 V, which motivates the near-threshold operation rather than the sub-threshold operation.

In first order, the propagation delay is determined by the drive current, the supply voltage and the load capacitor. This means there is not much room for improvement. A device with a larger drive current (e.g., a lower V_T) could improve propagation delay, as shown in Sect. 2.1.7. Overall, designers accept the speed reduction because it is so effective in decreasing the dynamic power. Very high-speed performance inevitably increases power consumption. Near-threshold operation results in a speed degradation, but is particularly good in realizing a lower leakage and dynamic power.

2.1.4 Leakage

While speed is determined by the absolute I_{on}, leakage power is determined by the absolute I_{off}. As I_{on} deteriorates at ultra-low voltage, so does I_{off}. The leakage

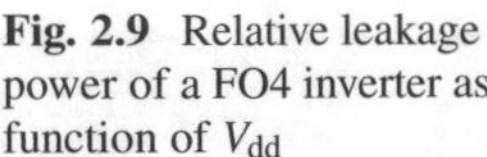

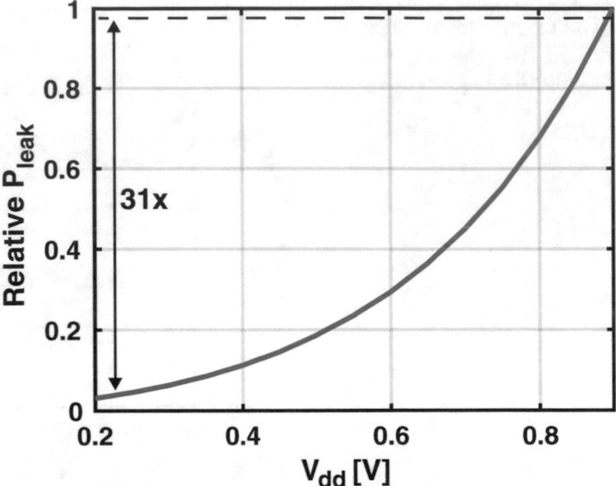

Fig. 2.9 Relative leakage power of a FO4 inverter as function of V_{dd}

power of the same FO4 inverter, simulated across the full supply voltage range, is shown in Fig. 2.9. Leakage power can be reduced severely by operating at ultra-low voltage. Comparing the delay degradation of Fig. 2.8 to the leakage power reduction demonstrates what was already shown in Sect. 2.1.2: the I_{on}/I_{off} reduces with decreasing supply voltage. The fraction of leakage power is thus expected to increase for near-threshold operation.

Although the leakage current and leakage power decrease, leakage energy typically increases with decreasing supply voltage. Leakage energy considers the leakage current integrated over the time it takes to complete the operation. Since lower V_{dd} operation results in heavily degraded speed, the integration time is significantly longer. This has a large impact, increasing the leakage energy.

2.1.5 *Process Variations*

CMOS transistors exhibit variations. Fluctuations of process and fabrication parameters, or just uncertainty on the process parameters themselves, result in performance differences. Inter-die variations classify variations on device parameters between different dice or wafers, while intra-die variations classify variations on device parameters between transistors on the same die. Inter-die variations are typically categorized into process corners, combining the effect of these variations in three categories for each device: *slow (S)*, *typical (T)* or *fast (F)*. The resulting process corners are then *TT*, *SS*, *FF*, *SF* and *FS*. Simulations in these five corners cover the entire inter-die variation space. Intra-die variations are represented by statistical distributions of varying device parameters. Simulation of these effects is done using Monte Carlo analysis. A typical example is variation of the threshold

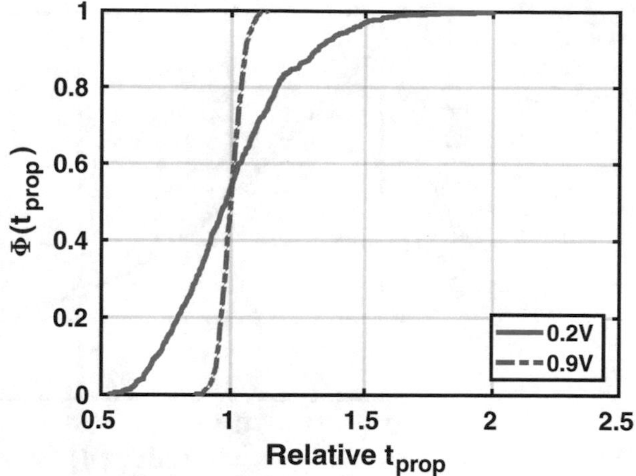

Fig. 2.10 Cumulative distribution function of FO4 inverter propagation delay intra-die variation for two cases: 0.2 V and 0.9 V. Simulations come from 500 Monte Carlo samples

voltage V_T, with, for example, a Gaussian distribution. As a result, I_{on} and I_{off} of the transistor will vary, as will the propagation delay and leakage power of a logic gate.

In the weak-inversion operating region, I_{on} and I_{off} depend exponentially on V_T. A slight variation in V_T can thus have a large impact on I_{on}, t_{prop} and P_{leak}. Figure 2.10 shows how the propagation delay of the FO4 inverter[1] varies under intra-die variation, simulated using Monte Carlo analysis. The cumulative distribution function for the near-threshold case (0.2 V) and the nominal case (0.9 V) is shown. Relative variation of the propagation delay is significantly larger for the near-threshold case. Figure 2.11 shows a similar effect: the propagation delay standard deviation (σ) when compared to the mean (μ) is significantly larger at ultra-low supply voltages. Inter-die variation sensitivity of the FO4 inverter propagation delay is shown in Fig. 2.12. Similar to intra-die variations, inter-die variations result in a relatively large shift in propagation delay.

The simulations just shown demonstrate how variable performance metrics are when operating in the near-threshold region. Both **intra- and inter-die variations result in unpredictability** of the propagation delay to the point where it is almost as likely to be double or halve the nominal delay. **Managing this unpredictability is the most demanding challenge in near-threshold operation**. Assuming the worst case performance variation is a very conservative way of tackling process variations. Variation-resilient considerations cross-cut this entire work, more specifically the rest of this chapter, Chaps. 4 and 6.

[1]From this point on simulations use the near-threshold optimized stacked nMOS inverter as shown in Fig. 2.20.

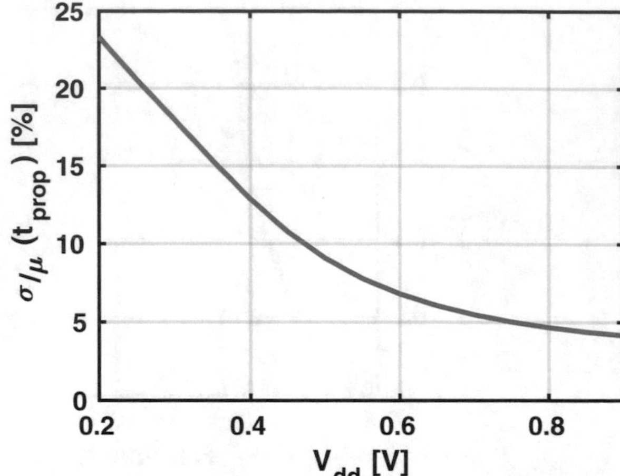

Fig. 2.11 $\sigma/\mu(t_{prop})$ of FO4 inverter intra-die variation as function of V_{dd}. Simulations come from 500 Monte Carlo samples

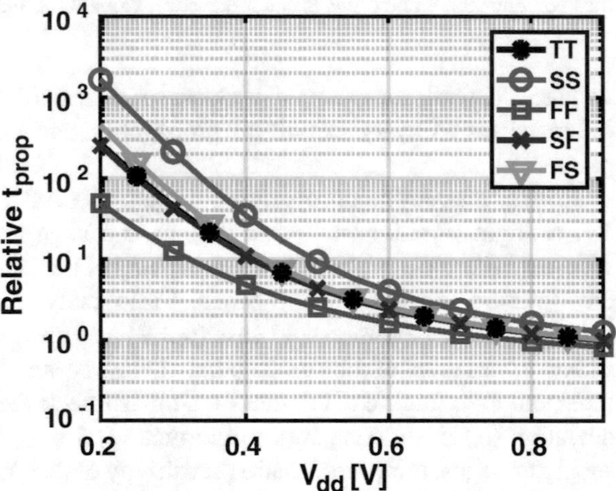

Fig. 2.12 Relative propagation of FO4 inverter as function of V_{dd} for different process corners or inter-die variations

2.1.6 Temperature

As shown in Eq. (2.1), V_T changes with temperature. Carrier mobility changes as well. Mobility and V_T both decrease with temperature [23, 30, 31]. For near-threshold operation, the effect on V_T is of more importance than at nominal voltage: the lower V_T results in a higher V_{gs}-V_T which increases I_{on}. The decreased mobility counteracts this effect, reducing I_{on}. These two effects counteract each other, resulting in a limited propagation delay temperature sensitivity. Figure 2.13 shows the effect of temperature on the propagation delay and leakage power of the FO4 inverter. Considering the propagation delay, the just described effect indeed holds:

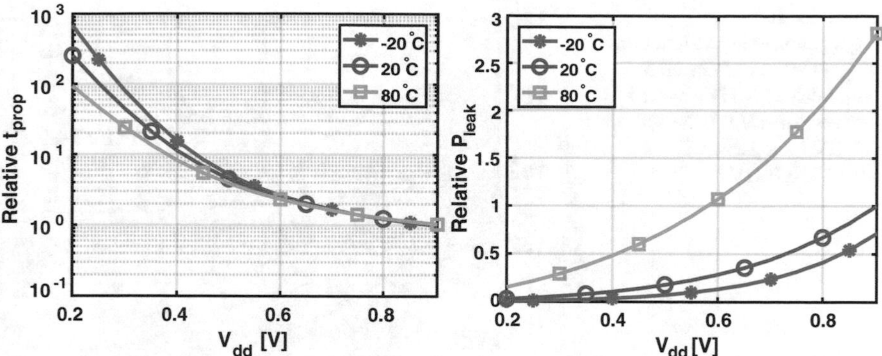

Fig. 2.13 Relative propagation delay for different ambient temperatures (left) and relative leakage power for different ambient temperatures (right) as function of V_{dd} for FO4 inverter

although t_{prop} does vary with temperature, the effect is limited. Extreme negative temperatures do degrade speed performance more severely [2]. Note that the lower V_T also increases I_{off}, resulting in a significant increase in leakage power, shown in Fig. 2.13.

Apart from process variations, ambient temperature variations are another important source of system unpredictability. The microcontroller systems developed in Chap. 4 demonstrate this by means of silicon measurements (see Sect. 4.4.6). In line with the FO4 inverter temperature variations, the microcontroller prototypes show limited temperature sensitivity in terms of speed, while the corresponding energy consumption does increase significantly.

2.1.7 V_T and Technology Flavour

To offer a more graceful trade-off between transistor size, delay and leakage power, recent CMOS technologies enable transistors implementation with different threshold voltages. Typical is the choice between three V_T's: high-V_T (HVT), regular-V_T (RVT) or low-V_T (LVT). In most technologies, these devices can be mixed and matched at will without significant overhead. To demonstrate the trade-off between the different V_T devices, Figs. 2.14 and 2.15 show the delay and leakage power of the same FO4 inverter. Since the most important difference between these cases is the V_T, the V_{gs}-V_T is higher for a lower-V_T device at a given supply voltage. This results in a higher I_{on} and I_{off}, thus a lower propagation delay and a higher leakage power.

The sensitivity to process variations is for a large part determined by the operating region of the transistors. For ultra-low voltage operation at the same V_{dd}, a higher V_T typically means the devices operate further in the weak-inversion region. As such, a higher V_T device demonstrates a higher sensitivity to process variations.

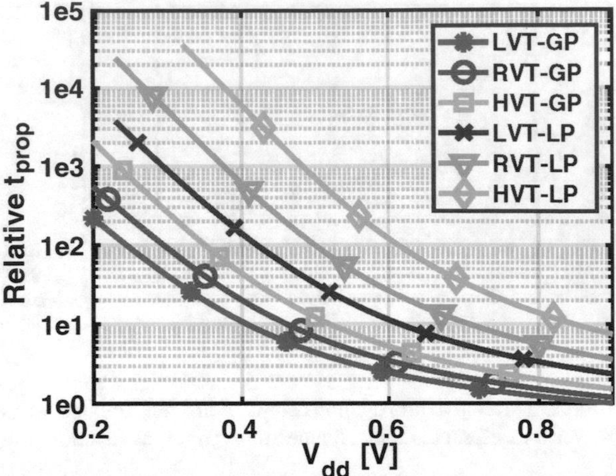

Fig. 2.14 Relative propagation delay as function of V_{dd} for the FO4 inverter implemented in six different devices: GP-LVT, GP-RVT, GP-HVT and LP-LVT, LP-RVT, LP-HVT

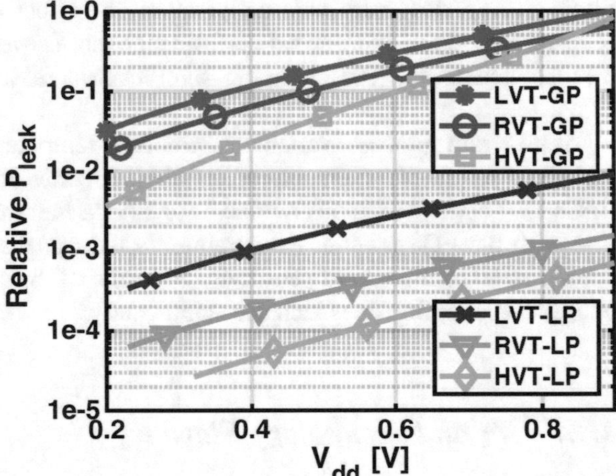

Fig. 2.15 Relative leakage power as function of V_{dd} for the FO4 inverter implemented in six different devices: GP-LVT, GP-RVT, GP-HVT and LP-LVT, LP-RVT, LP-HVT

Figure 2.16 shows this in a Monte Carlo simulation of the propagation delay of the FO4 inverter implemented in LVT, RVT and HVT devices. To achieve the same or better variation sensitivity, a higher V_T device needs to operate at a higher voltage, which increases both static and dynamic power.

Apart from different V_T devices, a single CMOS technology node often also comes in flavours. A typical distinction is the *general-purpose (GP)* flavour (something also called *high performance*) versus the *low-power (LP)* flavour. In contrast with the earlier discussed V_T differences, these flavours cannot be combined on a single chip. Similar to the V_T devices, the major difference between GP and LP is the V_T and by extension the available I_{on} for a given supply voltage. The distinction becomes clear when looking back at Figs. 2.14 and 2.15 which, apart from the GP LVT, RVT and HVT devices, also includes the LP LVT, RVT and HVT devices. Throughout this work, GP-LVT devices have been used for every implementation.

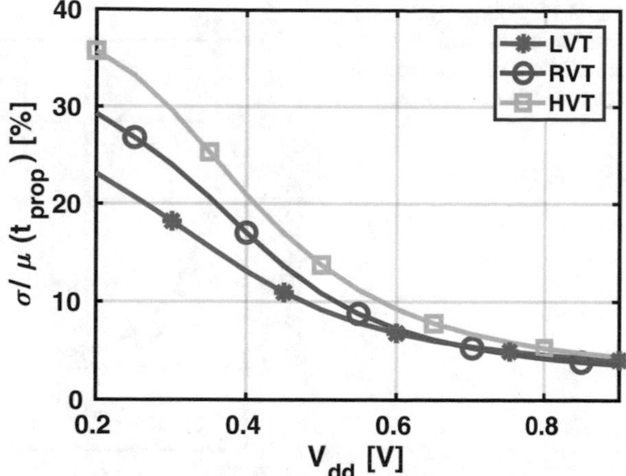

Fig. 2.16 $\sigma/\mu(t_{\text{prop}})$ of FO4 inverter intra-die variation as function of V_{dd} implemented in different V_T devices. Simulations come from 500 Monte Carlo samples.

Table 2.1 Microcontroller operating conditions, silicon measurement results and derived parameters

ARM Cortex-M0 core (Prototype 1)	
V_{dd}	350 mV
f_{clk}	12 MHz
P_{total}	170.00 μW
P_{static}	125.93 μW
P_{dynamic}	44.07 μW
C_{total}	383.37 pF
Activity α	7.23%

Apart from the higher I_{on} (and thus lower propagation delay) available because of this device, minimum energy operation is another major motivation affecting this choice. The next subsection demonstrates this by looking forward at one of the microcontroller prototypes.

2.1.8 Application in Prototypes

A single inverter demonstrates a significant delay degradation when operated in near-threshold operation. On the other hand, near-threshold operation also decreases leakage power substantially. These two effects suggest a trade-off. Especially since the analyses in this chapter always used the same constant size FO4 inverter. Load capacitance is thus fixed which means dynamic power improves significantly at ultra-low voltage.[2]

To further investigate this trade-off, we look forward to a typical operating point of the microcontroller implementation presented in Chap. 4. The ARM Cortex-M0 core silicon measurements achieve the results shown in Table 2.1. The operating

[2]As discussed in Chap. 1, dynamic power is proportional to V_{dd^2}, which is the main motivation for near-threshold operation.

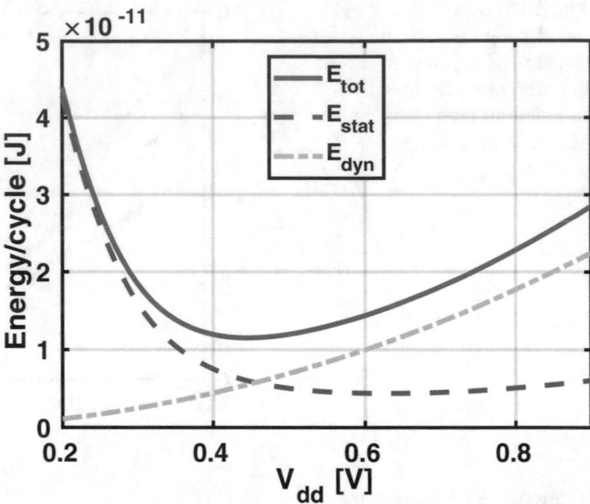

Fig. 2.17 Energy/cycle of the M0 core as predicted by the FO4 inverter propagation delay and leakage power

Table 2.2 Microcontroller operating conditions, silicon measurement results and derived parameters

ARM Cortex-M0 core (Prototype 1)		
	INV_{FO4} prediction	Measurement
V_{dd}	450 mV	440 mV
f_{clk}	36.1 MHz	31.2 MHz
E/cycle	11.52 pJ	16.07 pJ
Rel. I_{leak} contrib.	51.28%	68%

frequency and power consumption are measured results. The total capacitance is extracted from the layout through parasitic extraction, and the activity factor α is derived from the dynamic power, assuming it is directly determined by Eq. (1.1).

Now let's assume that the ARM Cortex-M0 core behaves exactly like the FO4 inverter discussed in this section. Prototype 1 of Chap. 4 is the best match for this, since it uses 40 nm LVT transistors only. Nevertheless, this assumption is very much simplified. The M0 core combines inverters, transmission gates and flip-flops to come to a complex system. If the measured speed and leakage power were to scale exactly like the LVT FO4 inverter, the energy results as shown in Fig. 2.17 would be achieved. The speed degradation, leakage power reduction and dynamic power reduction at near-threshold operating voltages indeed demonstrate a trade-off, leading to minimum energy consumption at 0.45 V. Static energy decreases with V_{dd} as a result of the faster clock frequency. Dynamic energy decreases quadratically due to its square relation to V_{dd}. Table 2.2 compares the predicted and actual minimum energy point. Considering the highly simplified assumption that the FO4 inverter exactly predicts the core performance, the modelling approach is fairly accurate.

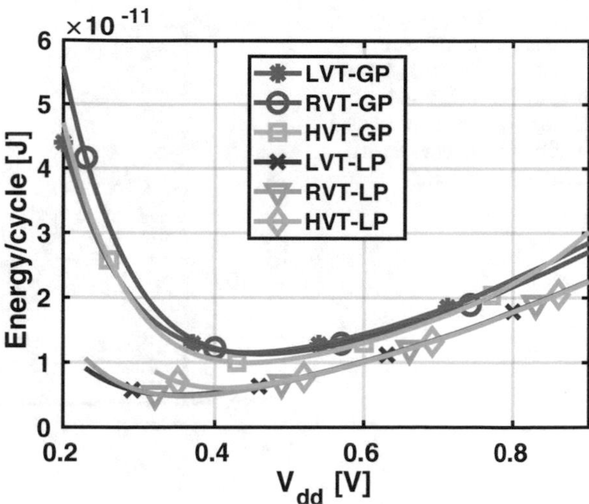

Fig. 2.18 Energy/cycle of the M0 core as predicted by the FO4 inverter propagation delay and leakage power, implemented in different V_T devices and technology flavours

Using this model, we can explore different V_T and technology flavours, in line with the GP-LP and LVT-RVT-HVT inverter simulation of Fig. 2.14. When doing this, gate capacitance typically does not change as device sizing is kept constant with changing V_T. The FO4 inverter simulations in these V_T's and technology flavours are used to predict ARM Cortex-M0 core energy consumption. Figure 2.18 shows how changing the threshold voltage has minimal impact on the energy per cycle. The minimum energy point demonstrates a nearly identical $V_{dd,MEP}$ and energy consumption for different V_T devices. The LP flavour shows a marginal energy reduction. Note that for the respective supply voltages where the LP flavour reaches minimum energy consumption, the devices are operating well below their V_T. This will impact functionality and variation sensitivity: the LP flavour FO4 inverter is not even functional at the lowest supply voltages under nominal conditions, as is visible in Fig. 2.18. Variation sensitivity will increase in line with the conclusions of Fig. 2.16.

Although the energy consumption hardly improves when changing V_T, the operating frequency does change significantly. A lower V_T yields a higher V_{gs}-V_T, which results in a larger drive current and thus a lower propagation delay of the inverter. This consideration is best represented when looking at the energy-delay product. Similar to the previous energy analysis, Fig. 2.19 shows the ARM Cortex-M0 core energy-delay product. When energy and operating frequency are combined in a single metric, the lowest V_T device (GP-LVT) achieves the best metric.

The combination of a similar energy consumption, a better variation sensitivity and a far better operating speed is the main motivation for the use of GP-LVT

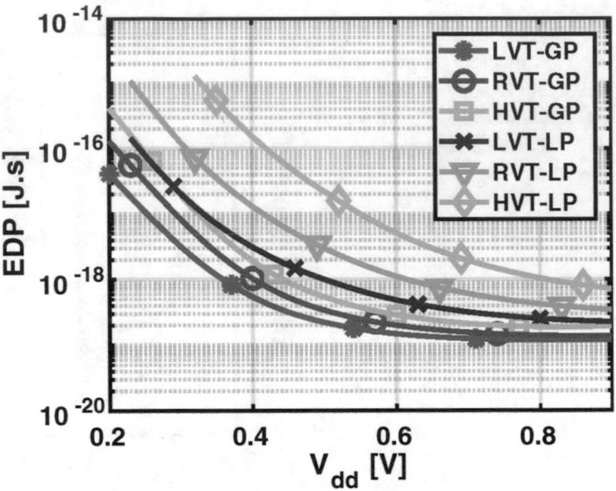

Fig. 2.19 Energy-delay product of the M0 core as predicted by the FO4 inverter propagation delay and leakage power, implemented in different V_T devices and technology flavours

devices in the rest of this work. Although the FO4 inverter based analysis simplifies some considerations, the silicon results of Chap. 4 are much in line with the conclusions presented here.

2.2 Logic Gates

To realize a near-threshold system, logic gates that are functional and variation resilient are the first step. The I_{on}–I_{off} analysis from Sect. 2.1 together with the FO4 inverter analysis provides some insights in how the typical standard CMOS topology behaves at ultra-low voltage.

2.2.1 Standard CMOS Gates

The I_{on} and I_{off} degradation at ultra-low voltage is most visible for the 40 nm technology pMOS devices used in this work. The most straight forward approach to overcome the forthcoming pMOS–nMOS imbalance is to enlarge the pMOS device. Figure 2.7 demonstrated this clearly, indicating a close to 11× larger pMOS is necessary to achieve the best possible noise margin. This strategy obviously has an effect on the gate area and input capacitance (and thus also dynamic energy). Although some area and dynamic power penalty can be tolerated if near-threshold operation is thereby enabled (and thus dynamic energy consumption is reduced), a better trade-off is possible.

Numerous circuit topologies are possible that deliver different trade-offs in area, propagation delay, leakage current and variation sensitivity. Reference [23] provides a thorough analysis on the most important subset. For the inverter (the most basic

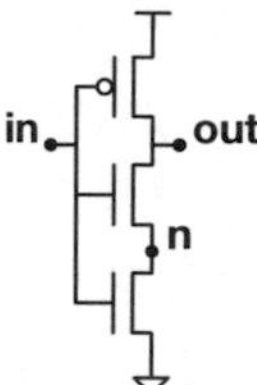

Fig. 2.20 Stacked nMOS inverter topology

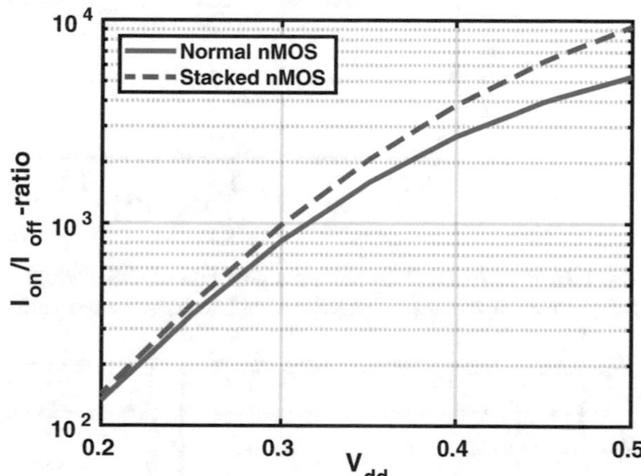

Fig. 2.21 Simulation on the effect of stacking on I_{on} and I_{off} across different V_{dd}

building block that provides regeneration), a stacked nMOS topology provides a lot of benefits [20, 23]. As visible in Fig. 2.20, the intermediary node n reduces the V_{ds} over each of the devices (DIBL decreases) and results in a negative V_{gs} and a positive V_{sb} for the top device. This makes it a very effective way of reducing the I_{off} when compared to non-stacked devices.

Apart from I_{off}, stacking also decreases I_{on}. With the difference of the intermediary node discussion earlier, a stacked topology is much like a single transistor with double the length. However, I_{off} decreases more than I_{on} which results in a better I_{on}/I_{off} ratio. Reference [20] clearly demonstrates this. Figure 2.21 shows a similar analysis across V_{dd} for 40 nm technology. When applying the stacking only for the nMOS pull-down, the decreased I_{on} has another advantage. It becomes easier to balance the pull-up/pull-down behaviour: the pMOS becomes relatively stronger due to the 'weakened' nMOS stack. As a result, optimal noise margin is achieved with a smaller pMOS device, clearly shown in Fig. 2.22.

The variation sensitivity of the stacked nMOS inverter is shown in Fig. 2.23. The graphic compares the stacked nMOS intra- and inter-die variation sensitivity to three instances of the normal inverter topology, sized for either equal area, equal propagation delay or equal noise margin. Intra-die variation is slightly better for the stacked nMOS case when compared to all others. For inter-die variation, the stacked nMOS inverter provides a good trade-off between area, noise margin and speed: the equal speed instance has very bad overall process corner performance; the equal area

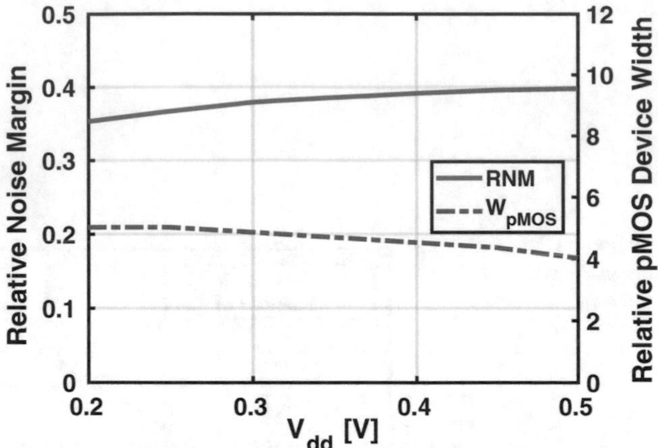

Fig. 2.22 Simulation of the noise margin of an nMOS stacked inverter for optimal relative pMOS sizing. Stacking relaxes the pMOS sizing constraint in comparison to Fig. 2.7

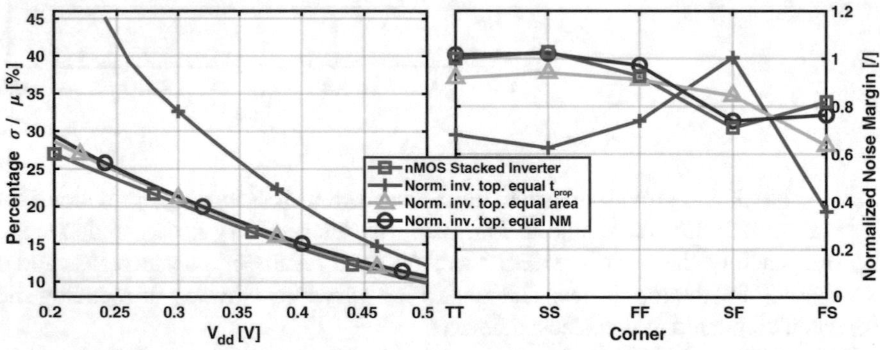

Fig. 2.23 Propagation delay variation of the stacked nMOS inverter under intra-die variation (left) and noise margin sensitivity to inter-die variation at 0.3 V (right). In both cases, the normal inverter topology sized for either equal area, equal propagation delay or equal noise margin is added for reference

instance also has degraded corner performance (especially for fast nMOS devices); and the equal noise margin instance requires significantly more area.

A more thorough analysis on the nMOS stacked inverter and the trade-offs it exposes can be found in [23]. In summary, the stacked nMOS inverter topology provides a good trade-off on several fronts:

- Better I_{on}/I_{off} ratios than a normal topology.
- Less area for similar noise margin and slightly reduced drive current.
- The best intra-die variation sensitivity in terms of speed performance.
- The best inter-die variation sensitivity trade-off.

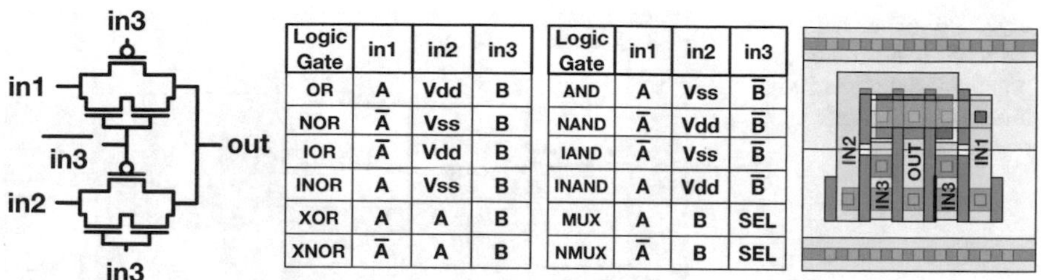

Logic Gate	in1	in2	in3
OR	A	Vdd	B
NOR	$\overline{A}$	Vss	B
IOR	$\overline{A}$	Vdd	B
INOR	A	Vss	B
XOR	A	A	B
XNOR	$\overline{A}$	A	B

Logic Gate	in1	in2	in3
AND	A	Vss	$\overline{B}$
NAND	$\overline{A}$	Vdd	$\overline{B}$
IAND	$\overline{A}$	Vss	$\overline{B}$
INAND	A	Vdd	$\overline{B}$
MUX	A	B	SEL
NMUX	$\overline{A}$	B	SEL

Fig. 2.24 Schematic of the differential transmission gate building block, where permutation of the inputs enables every logic function, including non-inverting functions

2.2.2 Transmission Gates

To realize near-threshold logic functions, a variety of logic topologies have been proposed. Apart from the conventional CMOS topology [3, 9, 11, 12, 14, 31], pseudo-nMOS logic [28], pass-transistor logic [1, 13], transmission gate logic [19, 20, 22] and dynamic leakage-suppression logic [10] are among the proposed topologies. The conventional CMOS topology is still the most popular approach: its ratioless static behaviour combined with high input–low output impedance makes it ideal to use in standard cells [17]. In this work, the transmission gate logic topology is used, in line with [19, 20, 22]. A thorough topology comparison and a detailed analysis on the transmission gate topology can be found in [23]. In this section, we briefly discuss a few key points and analyses.

A schematic overview of the transmission gate topology is shown in Fig. 2.24. Functionally, the transmission gate is a versatile building block. Every 2-input logic function is available with the same schematic by permutation of the input signals. Most of the following analyses thus hold for every logic function implemented using the transmission gate.

A key observation is the complementary pMOS–nMOS pair for both functional branches. From a pull-up/pull-down perspective, there is no need to balance the network, as both branches are identical. This is exceptionally useful when realizing more complex logic functions with >2 inputs like NOR3 or NOR4. Furthermore, the transmission gate R_{on} is the result of both the pMOS and the nMOS resistance in parallel. Upsizing the weaker pMOS device improves R_{on} only slightly, while it significantly increases gate capacitance [32]. Keeping the **transmission gate size small** thus not compromises functionality, while at the same time reducing input capacitance (both at the gate and drain of the devices). Remember, this was the case for the inverter discussed earlier, especially in 40 nm technology applied at ULV. Compared to pass-transistor logic, the transmission gate complementary structure can pass the full logic signal. To control the complementary pMOS–nMOS pair, a differential signal is required. Similar to [23], this work opts for a *fully differential approach*: every complementary signal pair is realized using a transmission gate, e.g., a TG-AND computes OUT, while a TG-NAND computes $\overline{\text{OUT}}$. This requires

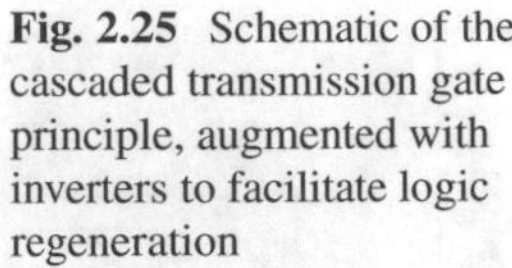

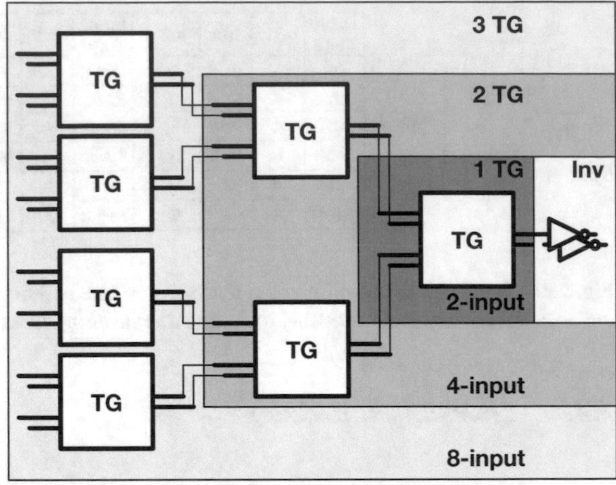

Fig. 2.25 Schematic of the cascaded transmission gate principle, augmented with inverters to facilitate logic regeneration

differential functionality as well as differential routing (something which is manageable, as will be demonstrated in Sect. 3.4.3.2). A different approach would be to create the complementary signal with a simple inverter.

Near-threshold application of the transmission gate topology requires some consideration, especially when considering the large difference between $I_{on,n}/I_{off,p}$ and $I_{on,p}/I_{off,n}$. To this end, the nMOS branch of the transmission gate uses stacked devices. This reduces the $I_{on,n}$ considerably, which improves the balance between the pull-up capability of the pMOS and the leakage of the pull-down nMOS. The transmission gate stacked nMOS improves (reduces) nMOS I_{off} behaviour, while nMOS stacking for the inverter improves (reduces) I_{on} behaviour [23].

As indicated in [23], the transmission gate does not regenerate the logic level. Even when the transmission gate is conducting, its R_{on} results in a signal loss at the output. For near-threshold operation, more than four consecutive transmission gates cannot be cascaded without compromising functionality in the 40 nm technology applied in this work. To cascade one or more consecutive transmission gates, this work adds an inverter after every functional transmission gate combination. A schematic representation is shown in Fig. 2.25. This work cascades up to three transmission gates, in line with the analysis presented in [23].

Differential operation introduces overhead, but also improves noise margin. Figure 2.26 shows that the relative noise margin of a differential transmission gate based inverter is higher than that of a single-ended inverter. Although the transmission gate based inverter is not used in this work, it is the same building block as the 2-input transmission gate based logic functions. Comparison of its noise margin with that of a single-ended inverter is a good metric to demonstrate the improvement due to differential operation.

A key property of the transmission gate topology is its reduced sensitivity to inter-die variations. When considering functionality, the *SF* and *FS* process corner are typically the most troublesome, since they heavily skew the pull-up/pull-down

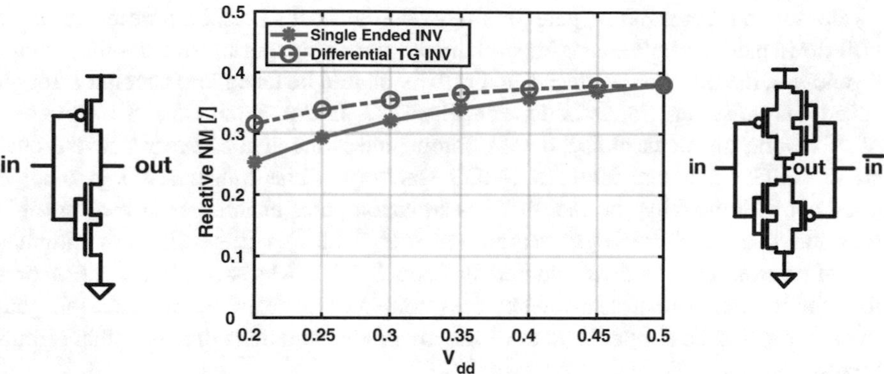

Fig. 2.26 Comparison of the relative noise margin of a single-ended stacked nMOS inverter and a differential transmission gate logic inverter as a function of V_{dd}

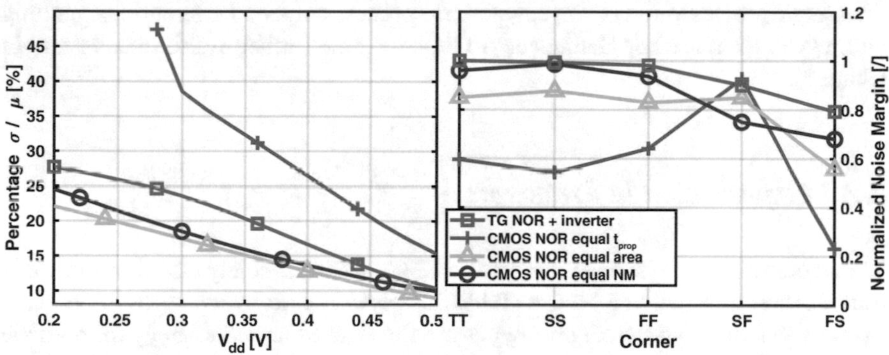

Fig. 2.27 Propagation delay variation of the transmission gate NOR under intra-die variation (left) and noise margin sensitivity to inter-die variation at 0.3 V (right). In both cases, the normal NOR topology sized for either equal area, equal propagation delay or equal noise margin is added for reference

balance. The transmission gate is inherently better, since both the pull-up and pull-down network have a pMOS–nMOS pair. Figure 2.27 evaluates both the intra- and inter-die variation sensitivity of the transmission gate topology implementing a 2-input NOR function. The conventional CMOS NOR gate is a critical gate since it requires pMOS stacking, something which is highly discouraged in the 40 nm CMOS technology of this work due to the relatively weak pMOS device in the near-threshold operating region. In Fig. 2.27, the CMOS NOR is designed for either equal propagation delay, equal area or equal noise margin. When considering intra-die variation, the small transistor sizing of the TG NOR slightly increases the σ/μ ratio when compared to the CMOS NOR. When considering inter-die variations, the pMOS–nMOS pair succeeds in enabling high noise margin for every process corner. Especially the *FS* corner demonstrates significantly better noise margin.

Because the transmission gate topology overcomes the need for a ratioed pull-up pull-down pair, per-gate area is kept relatively low. In comparing area to other circuit topologies, the ultra-low voltage functionality should be taken into account through speed and noise margin, as is done in Fig. 2.27. In the 40 nm CMOS technology of this work, the weak pMOS device compromises the area of every conventional static CMOS gate that requires pMOS stacking, while transmission gate logic does not. Additionally, the fact that transmission gates enable non-inverting logic functions can reduce total system area (see Sect. 3.4.2.4). Differential signalling has a limited overhead, as demonstrated in Sect. 3.4.3.2. Moreover, Table 4.6 shows that the total area of a near-threshold system with differential transmission gate logic appears to be similar to state-of-the-art implementations that use other circuit topologies.

In summary, the transmission gate topology allows relatively low area and low capacitive load logic gates that enable complex logic functions. Because of its pMOS–nMOS pairing, the transmission gate is particularly suited to counteract systematic process variations in asymmetric process corners. Differential signalling and regeneration are key challenges (to be overcome further in this chapter and in Chap. 3).

2.2.3 *Application in Prototypes*

The microcontroller systems presented in Chap. 4 consist solely out of inverter and transmission gate building blocks. Table 2.3 gives a short overview of the applied logic in each prototype. Both prototypes use the stacked nMOS strategy for both the inverter and the transmission gate.

Table 2.3 Application of near-threshold building blocks in microcontroller prototypes of Chap. 4

ARM Cortex-M0 core		
	Prototype 1	Prototype 2
Inverter	Stacked nMOS	
	40 nm gate length	$L = 40$ nm
	Relative sizing 5:1:1	Relative sizing 4:1:1
		$L = 60$ nm
		Relative sizing 6:1:1
Transmission gate	Stacked nMOS	
	Relative sizing 2:1:1	
	$L = 40$ nm	$L = 40$ nm
		$L = 60$ nm
Cascading	58% 1TG logic	20% 1TG logic
	37% 2TG logic	64% 2TG logic
	5% 3TG logic	16% 3TG logic

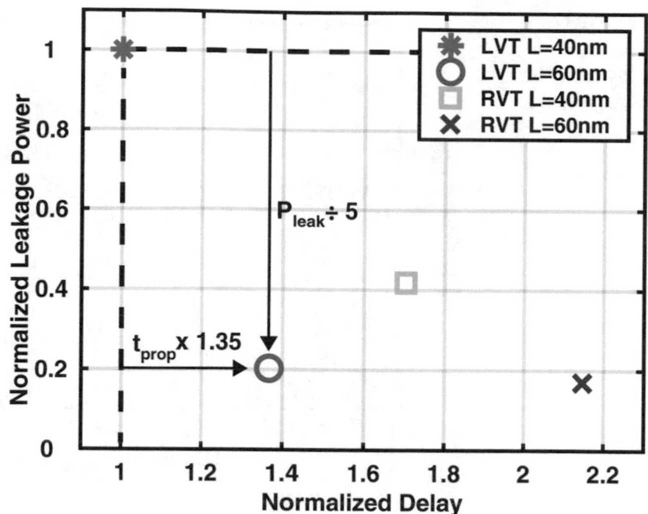

Fig. 2.28 Normalized leakage power and propagation delay of an nMOS stacked inverter with 40 nm and 60 nm gate length, low V_T and regular V_T devices

Looking forward to Chap. 4, one of the major improvements in prototype 2 was the use of a dual gate length library. The reverse short channel effect (RSCE) [29] in the applied 40 nm CMOS technology results in a significantly lower leakage power with a minimal increase in propagation delay. This trade-off is quantified in Fig. 2.28, which also compares with an RVT implementation. The RVT strategy actually increases leakage power, while propagation delay degrades further.

Both prototypes use up to three cascaded stages of transmission gates, in line with what was already shown in Fig. 2.25. Looking forward to the standard cell library developed in Chap. 3, the trade-off between propagation delay and leakage power when cascading multiple transmission gates becomes clear. Figure 2.29 compares, for both gate lengths, different logic implementations according to their propagation delay and leakage power. The comparison provides an intuitive look inside the logic synthesis considerations of the VLSI design flow. A simple 2-input logic gate (OR2 and AND2) has the lowest leakage power and propagation delay. The different gate length implementations exchange leakage power for speed. A 4-input logic gate (AND4, AOI and OR4) can be realized in two different ways: a single standard cell with two cascaded transmission gate stages, or three standard cells providing the same logic function (e.g., three AND2 gates can be combined into a single AND4 gate). The single cell performs better in terms of leakage power and propagation delay because it incorporates only one inverter. Note that in the case of the single cell 4-input gate, the result of an intermediary 2-input gate cannot be branched off to use elsewhere, since it is not driven by an inverter. A similar consideration is present for the 6-input logic gate (AND33OR2). Looking at this

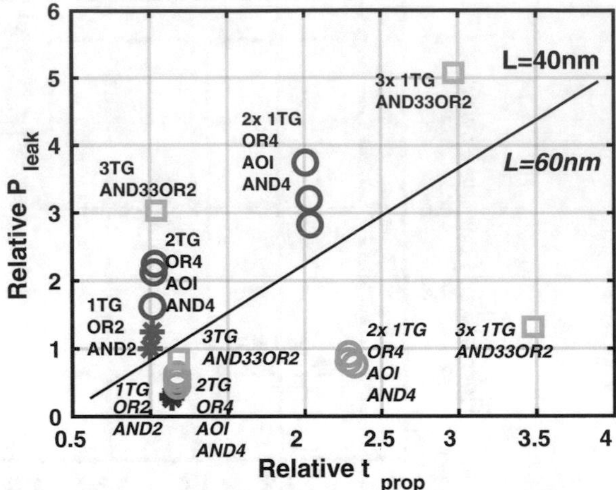

Fig. 2.29 Normalized leakage power and propagation delay for 40 nm and 60 nm gate length logic gates

small subset of logic gates coming from the standard cell libraries developed in Chap. 3, a large span in both leakage power and propagation delay is present, giving the logic synthesis tool enough degrees of freedom to implement a well performing design.

Finally, this chapter does not provide a detailed comparison between different logic gate topologies and their propagation delay, leakage power, area and variation sensitivity. Most of the works cited in the state-of-the-art comparison at the end of Chap. 4 employ conventional standard CMOS logic. These comparisons provide a system-level insight in the relative performance of the transmission gate logic used in this work.

2.3 Sequential Gates

Every system implemented in this work is synchronous and clock edge triggered. A single clock is distributed across the entire design, setting the pace for the system pipeline. The pipeline relies on flip-flops that sample the data-path on every rising clock edge, locking in the data to propagate to the subsequent logic path until the next rising clock edge. Near-threshold operation puts some additional constraints on the clock tree. Reliable distribution of the clock signal is discussed in more detail in Sects. 3.2.3 and 4.3.3. This section discusses the essential building blocks to enable synchronous operation: the latch and the flip-flop.

2.3.1 Latch

The latch is a level-sensitive sequential gate. The latch shown in Fig. 2.30 is transparent (passes input D to output Q) when the clock is low (clk is low, $\overline{clk}$ is high), and locks the output to the input value when the clock is high (clk is high, $\overline{clk}$ is low). To lock the output, the latch relies on a cross-coupled inverter pair where the feedback keeps the internal latch nodes stable. The transparent-lock operation is realized by disabling–enabling the feedback path. As the feedback is usually realized with inverters, a tri-state inverter is a convenient approach to enable–disable the feedback path and thus provides transparent-lock functionality.

This is especially convenient for near-threshold operation. Other approaches to latch operation rely on ratioed cross-coupled inverters, where the difference in drive strength is chosen to stabilize or overpower the latch node. At ULV, the enable–disable approach on the feedback path is much more variation resilient. In line with the discussion on logic gates in Sect. 2.2, the tri-state inverter is preferably not implemented in a conventional static CMOS topology because it requires a stacked pMOS circuit. As discussed in [23], a better approach is to use a transmission gate based tri-state inverter. In line with the discussion in Sect. 2.2, both the transmission gate and the inverter used in the latch employ nMOS stacking.

The differential transmission gate logic in Sect. 2.2 requires differential input and generates a differential output. Using differential sequential gates helps to have differential signals readily available. Because both the normal data and the

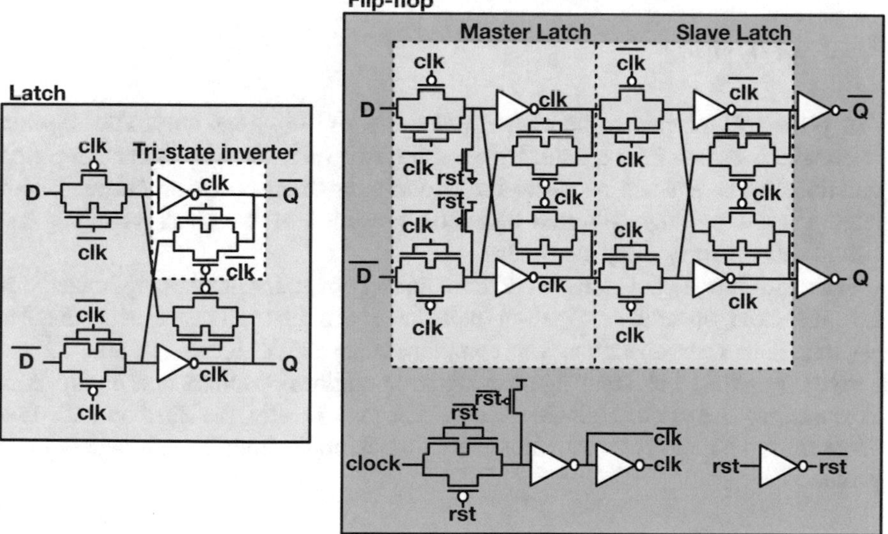

Fig. 2.30 Transistor implementation of the positive edge triggered flip-flop (right) used in this work, consisting of two level-sensitive latches (left)

complementary data are always available in the latch, the differential operation is easily implemented.

Latch-based pipelines have been used extensively for near-threshold operation. Reference [22] implements a ULV-enabled full JPEG encoder in a two-phase latched based pipeline. More recently, latch-based pipelines have been used in low or ultra-low voltage designs combined with timing error detection (see Chap. 5). Reference [7] compares extensively between flip-flop, two-phase latch-based and pulsed latch-based implementations.

Latch-based pipelines are particularly variation resilient because of their time borrowing capability [23]. When intra-die variation results in a propagation delay that exceeds the length of the locked phase of the latch, the transparent phase allows the data to propagate through the latch without compromising functionality. A major drawback is that this time borrowing functionality is rarely leveraged during static timing analysis in the VLSI design flow. Latches are typically analysed without considering the transparent phase, making sure the data arrives within the locked phase of the latch. Another challenge is the differential clock tree in two-phase latch-based design. To operate a two-phase latch-based pipeline, each clock phase is distributed separately having a non-overlapping phase to make sure that two subsequent latches are not transparent at the same time. This results in a double clock tree distribution which increases the clock energy and area.

Because of these challenges, most sequential designs use flip-flop based pipelines. A flip-flop combines two latches, which makes it edge triggered and much easier to use during static timing analysis.

2.3.2 Flip-Flop

The prototypes discussed throughout this work use flip-flops only. The flip-flop schematic is shown in Fig. 2.30. By combing two level-sensitive latches, flip-flop functionality is enabled: back-and-forth lock-transparent operation between both latches results in edge triggered operation governed by the clock, changing the output data on every rising clock edge.

The flip-flop used in this work combines two of the latches optimized for near-threshold operation. A pull-up/pull-down pair in the master latch enables asynchronous reset operation. The complementary clock signals clk and $\overline{clk}$ are derived from the clock input signal using two additional inverters in each flip-flop. This enables the use of a single-ended clock tree. Finally, the flip-flop adds two inverters at the output to isolate the internal latch nodes from the subsequent logic paths.

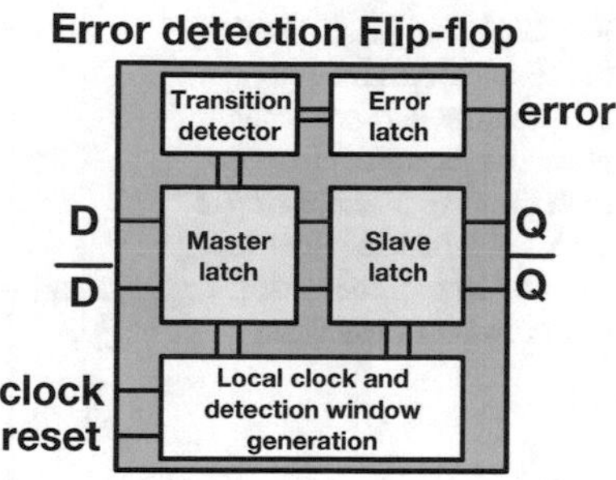

Fig. 2.31 Block diagram of the error detection flip-flop, further discussed in Chap. 6

2.3.2.1 Error Detection Flip-Flop

Chapters 5 and 6 discuss margin reduction by means of error detection and correction. The error detection flip-flop is the key building block that enables the augmented functionality of these flip-flops. A typical aspect of sequential pipelines is that they add margin to the clock period to make sure every logic path can fully compute in that time. In near-threshold operation enabled pipelines, this results in a large overhead due to margining. Error detection flip-flops (see Fig. 2.31) provide system-level feedback regarding the margined operation of the circuit. Applying these flip-flops in a DVS-enabled feedback system thus results in significantly smaller overhead due to margins. A thorough analysis and system implementation regarding error detection and correction, error detection flip-flops and system-level error feedback is given in Chaps. 5 and 6.

2.3.3 Application in Prototypes

The two prototypes presented in Chap. 4 both use positive clock edge triggered flip-flops to implement a pipelined microcontroller system. The ARM Cortex-M0 core of those systems has 841 flip-flops controlling three distinct pipeline stages. Looking at the flip-flop building block in Fig. 2.30, it combines much more functionality than the simple logic gates discussed in Sect. 2.2. The number of inverters incorporated in each flip-flop is high, resulting in a significant leakage power per flip-flop. Figure 2.32 shows the leakage contribution of flip-flops to the total leakage power of both prototypes developed in Chap. 4. In both prototypes, the sequential logic contributes close to 30% of the total leakage power.

Out of all the flip-flops in the microcontroller system, Chap. 6 converts 5.7% to error detection flip-flops. Because of this, the microcontroller becomes timing error aware. Worst case margins adhering to *SS* process corners can be compensated through dynamic voltage scaling, reducing the energy consumption of a *TT* silicon implementation with 75% when compared to operating it at the worst case *SS* operating point.

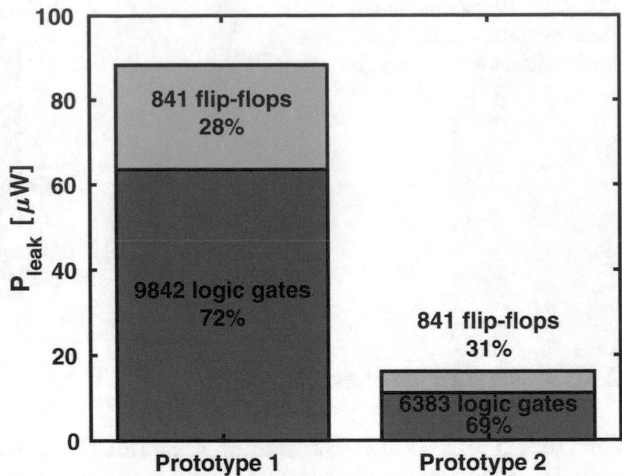

Fig. 2.32 Bar graph showing the absolute and relative contribution to the leakage power of sequential logic to both prototypes, simulated at 0.3 V. Flip-flops contribute close to 30% of the leakage power in both prototypes

2.4 Architecture

A near-threshold system aims to operate at the minimum energy point. At the minimum energy point, both static and dynamic energy contribute to the energy consumption almost equally. As long as the system is functional across a wide enough voltage range, sweeping the supply voltage while operating at the maximum operating frequency will always result in an energy minimum. For a given system, this is the most energy efficient point to operate the system. Looking at the main energy contributors, static energy increases when the clock frequency decreases (the leakage current is integrated over the clock period), while the dynamic energy increases with higher activity. The architecture of a digital system has an impact on both and can thus influence the minimum energy point voltage, operating frequency and energy consumption.

2.4.1 Logic Depth and Pipelining Depth

Logic depth and pipelining depth are two architectural aspects that influence activity and operating frequency. A simple example is shown in Fig. 2.33. For the same logic path, three different pipeline depths are shown. Having more pipeline stages results in a higher relative activity and a shorter clock period. For systems subjected to the same data, the static energy decreases with decreasing clock period, while the dynamic energy increases with activity. These two effects both result in a minimum energy point at a lower V_{dd}. The lower supply voltage, in its turn, reduces dynamic energy. Generally, a near-threshold system thus benefits from a fast operating speed and a high activity.

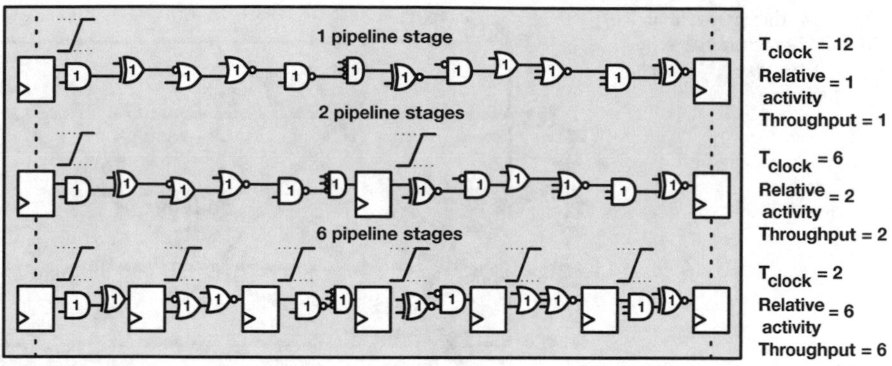

Fig. 2.33 For the same system, different pipeline depths result in different activities and maximum clock periods

The JPEG encoder in [22], the multiply–accumulate in [21, 26] and the adder in [19] all employ a deep pipeline, having a latch every two or three logic gates. The result is a highly efficient data-path block with a high throughput, operating frequency and activity.[3] Near-threshold operation is ideal for these type of systems: because of the high activity and fast operating frequency, the minimum energy point is located at a very low supply voltage. [6] elaborates on ultra-low voltage operation of deep pipelines.

For a general digital system, this level of architectural decisions is not always available. The high throughput of a deeply pipelined digital system also requires constant data input to operate efficiently. Sparsely triggered sensor processing systems can have a relatively high *idle* time. Other architectural optimizations such as clock and power gating can be used to increase the relative activity of the circuit and decrease the energy consumption of idle (sub-)systems. The impact of these optimizations is discussed in Sect. 3.2.1.

2.4.2 Application in Prototypes

The microcontroller systems implemented in Chap. 4 are architecturally defined by their instruction set and their industry-supported ecosystem. During simulations and silicon measurements, the system activity is determined by the C-program running on the microcontroller. Throughout all simulations and measurements, the same identical C-code benchmark is used (see Sect. 4.2.4) to trigger the circuit, resulting in the same activity rate for each measurement point.

[3]The actual activity rate depends as much on the incoming data as on the amount of pipeline stages.

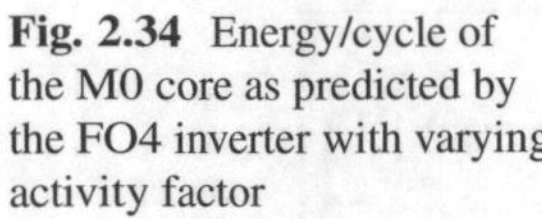

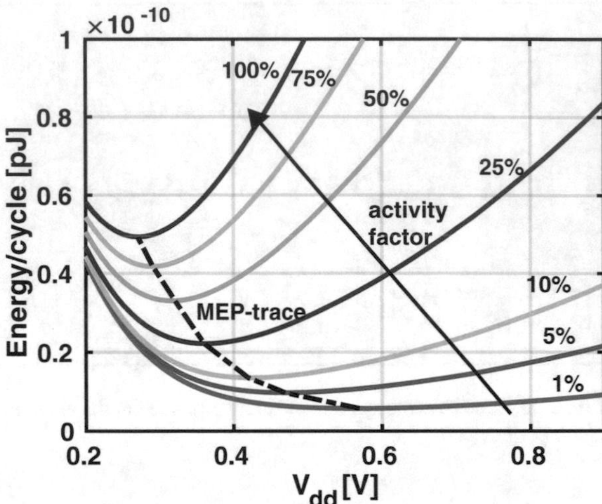

Fig. 2.34 Energy/cycle of the M0 core as predicted by the FO4 inverter with varying activity factor

Different in-field applications of the microcontroller system will result in wide spectrum of activity rates. Each activity rate will result in a different minimum energy point supply voltage, speed and energy consumption. Figure 2.34 demonstrates how the microcontroller system energy/cycle varies if the activity rate changes. With higher activity rate, the minimum energy point shifts to a lower supply voltage and a higher energy consumption. The higher activity increases dynamic energy which then becomes relatively larger, skewing the minimum energy trade-off.

2.5 CMOS Technology Advancements

The analyses and reported performance presented in this chapter are particular to the 40 nm CMOS technology in this work. Although the conclusions presented here are as generic as possible, some of them change when technology parameters such as sub-threshold slope and relative pMOS strength change. An overall observation is that the relative pMOS strength improves significantly in most technologies below 40 nm CMOS. Recent advanced technologies also improve the sub-threshold slope. In [32] the sub-threshold slope (S) is defined as the amount of gate voltage necessary to increase the current by an order of magnitude in the weak-inversion region. It can be calculated as $n \cdot v_T \cdot \ln(10)$, where n is a process-dependent parameter and v_T is the thermal voltage (26 mV at room temperature). For a typical CMOS process $n = 1.5$, resolving S to 90 mV/decade [27, 32]. In the ideal case $n = 1$, which means $S = 60\,\text{mV/decade}$. The lower the sub-threshold slope, the better the I_{on}/I_{off} behaviour of the technology. Near-threshold operation can benefit significantly from technologies with a lower sub-threshold slope [8].

2.5.1 Fully Depleted Silicon-on-Insulator Technology

A recent technological development of particular interest for ultra-low voltage operation is 28 nm ultra-thin buried box fully depleted silicon-on-insulator (UTBB FD-SOI). A cross section of a UTBB FD-SOI transistor is shown in Fig. 2.35. The transistors have a ultra-thin (25 nm) buried oxide to dielectrically isolate the transistor from the substrate. This results in reduced drain/source-substrate parasitic capacitance, a lower leakage current and latch-up immunity [5]. This fully depleted technology does not require doping implants in the channel which means the channel does not suffer from random dopant fluctuations (RDF). This improves the device variability, something which is highly encouraged for ultra-low voltage operation. This SOI technology has been reported to have a sub-threshold slope of 85 mV/decade [32], thus improving on bulk CMOS technology. Because of the buried oxide, the transistor body is isolated from the source and drain, which allows body biasing over a wide range without inflicting latch-up. This way, body biasing can be used to modulate the device V_T with up to 85 mV/V body biasing. Scaling the V_T of a digital system in such a way (at run-time) can enable a much wider spectrum in speed–energy combinations.

A key difference between FD-SOI technology and the 40 nm CMOS technology used in this work is the relatively good pMOS device properties at ultra-low voltage. The pMOS–nMOS ratio to achieve optimal noise margin at ultra-low voltage is limited to 2 [26]. A possible use of body biasing could be to eliminate this imbalance completely by asymmetrically biasing the pMOS and nMOS devices, resulting in the best possible noise margin and thus very good ultra-low voltage variation resilience with equal pMOS and nMOS device sizes.

Because of the better sub-threshold slope, good variability and wide body biasing range, FD-SOI technology is widely being used for ultra-low voltage digital systems. Section 2.5.3 discusses a prototype implementation of a 16-bit multiply accumulate in FD-SOI technology.

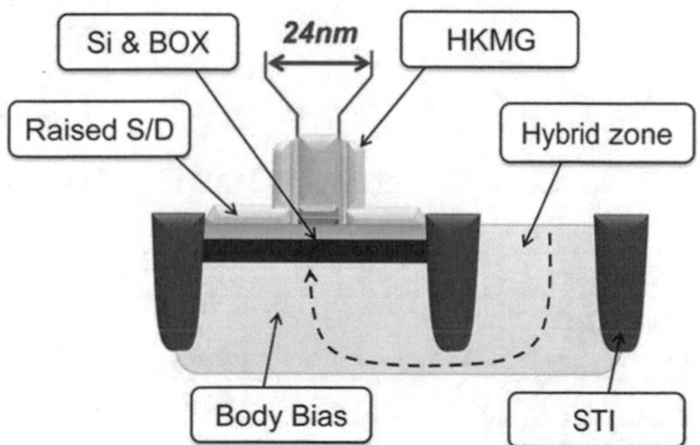

Fig. 2.35 Cross section of the UTBB FD-SOI transistor device, taken from [5]

2.5.2 FinFET CMOS Technology

An even more advanced recent CMOS technology development is the use of finFET transistors. Similar to SOI technology, the sub-threshold slope is reduced. In the case of finFET devices, this stems from the fact that the gate surrounds the channel on more sides, typically three, hence also called tri-gate CMOS. Again, it allows a better I_{on}/I_{off} behaviour since the fin-like gate structure allows to turn off the transistor more abruptly [32]. Reference [8] assesses the properties of multi-gate MOSFET devices further, as does [16]. Reference [15] is a recent ultra-low voltage microcontroller implementation in 14 nm tri-gate CMOS that is included in the state-of-the-art comparison in Chap. 4.

2.5.3 Application in Prototypes

Experiments in 28 nm FD-SOI technology resulted in the implementation and measurements of a 16-bit multiply–accumulate block (MAC) [26]. The chip micrograph is shown in Fig. 2.36. As shown in Table 2.4, it demonstrates an 8× energy-delay improvement for the same block implemented in 90 nm bulk CMOS technology,

Fig. 2.36 Chip micrograph of the 16-bit MAC implementation in 28 nm UTBB FD-SOI

Table 2.4 Comparison of a 16-bit MAC block in 90 nm bulk CMOS [21] and 28 nm UTBB FD-SOI [26]

Technology	90 nm bulk [21]	28 nm UTBB FD-SOI [26]				
Operation mode	MEP	$V_{dd,min}$	MEP	MEP @ no V_{BB}	MEP @ $V_{BB,min}$	MEP @ $V_{BB,max}$
V_{dd}[V]	0.19	0.21	0.25	0.29	0.33	0.29
V_{BB}[V]	/	1.0	0.5	0.0	−0.3	1.5
f_{clk} [MHz]	10	8.5	35	25	26	147
P_{leak}[μW]	3.90	5.67	2.13	0.95	0.65	33.52
E_{tot}[pJ]	0.87	1.04	0.17	0.28	0.32	0.42

even without body biasing. Using the body bias, different operating points can be achieved. The minimum energy point can be improved with 40% through a slight body voltage. A moderate body voltage can improve the minimum functional supply voltage by 28%. Speed can be improved almost 6× using a high body voltage, while leakage power can be reduced with 32% using a negative body voltage. More details on the variation resilience and other improvements this technology brings can be found in [26]. Finally, note that some of the works reported in the state-of-the-art comparisons in Chap. 4 use FD-SOI technology, demonstrating its good properties for ultra-low voltage operation in a microcontroller implementation.

2.6 Conclusion

This chapter laid out the ground work for near-threshold operation. By looking at the most basic transistor properties and their influence on the performance of a simple FO4 inverter, the challenges for near-threshold operation become quite clear. Enabling **high-speed performance** is a first major challenge. The inherently low drive current at ultra-low voltage reduces operating speed by two orders of magnitude or more when comparing to nominal supply voltage operation. A low-area implementation results in a smaller load capacitance which improves speed. This is realized through the use of a **stacked nMOS topology** in the inverter and the **inherently well-balanced differential transmission gate**. Furthermore, the use of **LVT devices** in a **general-purpose technology flavour** gives the maximum possible drive current for the same supply voltage.

When operation in the near-threshold region is enabled, the reduction in dynamic energy results in a relatively higher leakage power. When the speed performance then degrades, the **leakage energy** increases proportionally. The **stacked nMOS topology** improves I_{off} of nMOS and pMOS, while **cascading multiple transmission gates** enables more functionality for only slightly higher leakage power. Furthermore, a **60 nm gate length** logic gate proves to be a better trade-off in terms of speed and leakage power. Both 40 nm and 60 nm devices were used to implement the microcontroller systems further on in this work.

Near-threshold operation is highly susceptible to **process variations**. It shares a lot of properties with the weak-inversion region, making it hard to guarantee functionality and speed performance across intra- and inter-die process variations. The **stacked nMOS topology** and **differential transmission gate topology** improve influence of intra- and inter-die process variations to the best of their ability. **Timing error detection flip-flops** were briefly introduced in this chapter, as they allow optimal removal of the voltage margins that are usually applied to compensate for process variations.

The transmission gate building blocks conflict with some parts of the **common VLSI design flow**. The use of an **inverter at the end of every transmission gate chain** combined with **clock edge triggered flip-flops** removes much of the obstacles preventing a full-blown VLSI design approach. The remaining obstacles (differential signalling and ULV-aware physical implementation) are tackled in Chap. 3. The standard cell approach developed in that chapter opens up a wide range of **architectural optimization** techniques that improve near-threshold operation. Logic synthesis, timing analysis, power analysis and verification improve the **predictability** of larger digital systems such as the ARM Cortex-M0 microcontroller and enable **static vs. dynamic energy considerations** that influence minimum energy consumption. The conclusions presented in this chapter were published in [24, 25].

References

1. Alarcón, L.P., Liu, T.T., Pierson, M.D., Rabaey, J.M.: Exploring very low-energy logic: a case study. J. Low Power Electron. **3**, 223–233 (2007)
2. Bol, D., Hocquet, C., Flandre, D., Legat, J.D.: The detrimental impact of negative Celsius temperature on ultra-low-voltage CMOS logic. In: 36th IEEE European Solid-State Circuits Conference (ESSCIRC), pp. 522–525. IEEE, Piscataway (2010)
3. Bol, D., De Vos, J., Hocquet, C., Botman, F., Durvaux, F., Boyd, S., Flandre, D., Legat, J.D.: SleepWalker: a 25-MHz 0.4-V sub-mm2 7-uW/MHz microcontroller in 65-nm LP/GP CMOS for low-carbon wireless sensor nodes. IEEE J. Solid State Circuits **48**(1), 20–32 (2013)
4. Dreslinski, R.G., Wieckowski, M., Blaauw, D., Sylvester, D., Mudge, T.: Near-threshold computing: reclaiming Moore's law through energy efficient integrated circuits. Proc. IEEE **98**(2), 253–266 (2010)
5. Jacquet, D., Hasbani, F., Flatresse, P., Wilson, R., Arnaud, F., Cesana, G., Di Gilio, T., Lecocq, C., Roy, T., Chhabra, A., Grover, C., Minez, O., Uginet, J., Durieu, G., Adobati, C., Casalotto, D., Nyer, F., Menut, P., Cathelin, A., Vongsavady, I., Magarshack, P.: A 3 GHz dual core processor ARM Cortex TM -A9 in 28 nm UTBB FD-SOI CMOS with ultra-wide voltage range and energy efficiency optimization. IEEE J. Solid State Circuits **49**(4), 812–826 (2014)
6. Jeon, D., Seok, M., Chakrabarti, C., Blaauw, D., Sylvester, D.: A super-pipelined energy efficient subthreshold 240 MS/s FFT core in 65 nm CMOS. IEEE J. Solid State Circuits **47**(1), 23–34 (2012)
7. Jin, W., Kim, S., He, W., Mao, Z., Seok, M.: In situ error detection techniques in ultralow voltage pipelines: analysis and optimizations. IEEE Trans. Very Large Scale Integr. VLSI Syst. **25**(3), 1032–1043 (2017)
8. Kim, J.J., Roy, K.: Double gate-MOSFET subthreshold circuit for ultralow power applications. IEEE Trans. Electron Devices **51**(9), 1468–1474 (2004)
9. Kwong, J., Ramadass, Y.K., Verma, N., Chandrakasan, A.P.: A 65 nm sub-Vt microcontroller with integrated SRAM and switched capacitor DC-DC converter. IEEE J. Solid State Circuits **44**(1), 115–126 (2009)
10. Lim, W., Lee, I., Sylvester, D., Blaauw, D.: Batteryless sub-nW cortex-M0+ processor with dynamic leakage-suppression logic. In: IEEE International Solid-State Circuits Conference Digest of Technical Papers (ISSCC), pp. 1–3. IEEE, Piscataway (2015)
11. Luetkemeier, S., Jungeblut, T., Porrmann, M., Rueckert, U.: A 200mV 32b subthreshold processor with adaptive supply voltage control. In: IEEE International Solid-State Circuits Conference Digest of Technical Papers (ISSCC), pp. 484–486. IEEE, Piscataway (2012)
12. Mäkipää, J., Turnquist, M.J., Laulainen, E., Koskinen, L.: Timing-error detection design considerations in subthreshold: an 8-bit microprocessor in 65 nm CMOS. J. Low Power Electr. Appl. **2**(2), 180–196 (2012)
13. Markovic, D., Wang, C., Alarcon, L., Tsung-Te Liu, Rabaey, J.: Ultralow-power design in near-threshold region. Proc. IEEE **98**(2), 237–252 (2010)

14. Myers, J., Savanth, A., Howard, D., Gaddh, R., Prabhat, P., Flynn, D.: An 80nW retention 11.7pJ/cycle active subthreshold ARM Cortex-M0+ subsystem in 65 nm CMOS for WSN applications. In: IEEE International Solid-State Circuits Conference Digest of Technical Papers (ISSCC), pp. 1–3. IEEE, Piscataway (2015)
15. Paul, S., Honkote, V., Kim, R.G., Majumder, T., Aseron, P.A., Grossnickle, V., Sankman, R., Mallik, D., Wang, T., Vangal, S., Tschanz, J.W., De, V.: A sub-cm3 energy-harvesting stacked wireless sensor node featuring a near-threshold voltage IA-32 microcontroller in 14-nm tri-gate CMOS for always-ON always-sensing applications. IEEE J. Solid State Circuits **52**(4), 961–971 (2017)
16. Pinckney, N., Shifren, L., Cline, B., Sinha, S., Jeloka, S., Dreslinski, R.G., Mudge, T., Sylvester, D., Blaauw, D.: Near-threshold computing in FinFET technologies: opportunities for improved voltage scalability. In: Proceedings of the 53rd Annual Design Automation Conference (DAC), pp. 1–6. IEEE, New York (2016)
17. Rabaey, J., Chandrakasan, A.P., Nikolic, B.: Digital Integrated Circuits: A Design Perspective, 2nd edn. Pearson Education Inc., Hoboken (2003)
18. Razavi, B.: Design of Analog CMOS Integrated Circuits, 2nd edn. McGraw Hill Education, New York (2017)
19. Reynders, N., Dehaene, W.: A 190mV supply, 10MHz, 90 nm CMOS, pipelined sub-threshold adder using variation-resilient circuit techniques. In: IEEE Asian Solid-State Circuits Conference (A-SSCC), pp. 113–116. IEEE, Piscataway (2011)
20. Reynders, N., Dehaene, W.: Variation-resilient building blocks for ultra-low-energy sub-threshold design. IEEE Trans. Circuits Syst. Express Briefs **59**(12), 898–902 (2012)
21. Reynders, N., Dehaene, W.: Variation-resilient sub-threshold circuit solutions for ultra-low-power digital signal processors with 10MHz clock frequency. In: 38th IEEE European Solid-State Circuits Conference (ESSCIRC), pp. 474–477. IEEE, Piscataway (2012)
22. Reynders, N., Dehaene, W.: A 210mV 5MHz variation-resilient near-threshold JPEG encoder in 40 nm CMOS. In: IEEE International Solid-State Circuits Conference Digest of Technical Papers (ISSCC), pp. 456–457. IEEE, Piscataway (2014)
23. Reynders, N., Dehaene, W.: Ultra-Low-Voltage Design of Energy-Efficient Digital Circuits. Springer, Leuven (2015)
24. Reyserhove, H., Dehaene, W.: A 16.07pJ/cycle 31MHz fully differential transmission gate logic ARM Cortex M0 core in 40 nm CMOS. In: 42nd IEEE European Solid-State Circuits Conference (ESSCIRC), pp. 257–260. IEEE, Piscataway (2016)
25. Reyserhove, H., Dehaene, W.: A differential transmission gate design flow for minimum energy sub-10-pJ/cycle ARM Cortex-M0 MCUs. IEEE J. Solid State Circuits **52**(7), 1904–1914 (2017)
26. Reyserhove, H., Reynders, N., Dehaene, W.: Ultra-low voltage datapath blocks in 28 nm UTBB FD-SOI. In: IEEE Asian Solid-State Circuits Conference (A-SSCC), pp. 49–52. IEEE, Piscataway (2014)
27. Roy, K., Mukhopadhyay, S., Mahmoodi-Meimand, H.: Leakage current mechanisms and leakage reduction techniques in deep-submicrometer CMOS circuits. Proc. IEEE **91**(2), 305–327 (2003)
28. Soeleman, H., Roy, K.: Ultra-low power digital subthreshold logic circuits. In: Proceedings of the International Symposium on Low Power Electronics and Design (ISLPED), pp. 94–96. ACM Press, New York (1999)
29. Tae-Hyoung Kim, Keane, J., Hanyong Eom, Kim, C.: Utilizing reverse short-channel effect for optimal subthreshold circuit design. IEEE Trans. Very Large Scale Integr. VLSI Syst. **15**(7), 821–829 (2007)
30. Tsividis, Y., McAndrew, C.: Operation and Modeling of the MOS Transistor, 3rd edn. Oxford University Press, London (2011)
31. Wang, A., Calhoun, B.H., Chandrakasan, A.P.: Sub-threshold Design for Ultra Low Power Systems. Springer, New York (2006)
32. Weste, N., Harris, D.: CMOS VLSI Design: A Circuits and Systems Perspective, 4th edn. Addison-Wesley Publishing, Boston (2010)

Chapter 3
Efficient VLSI Design Flow

Abstract The efficient design flow presented in this chapter is a crucial aspect of this work. Full custom design approaches succeed in enabling ultra-low energy operation, but are too time intensive for large-scale digital designs. Commercial standard cell libraries and very-large-scale-integration (VLSI) design flows enable a fast design methodology for large-scale digital designs, but fail to provide ultra-low energy operation. This chapter unifies these two. Chapter 2 presented a logic library which has proven to be performant under ultra-low-voltage variation-sensitive conditions. This chapter takes that logic library to the next level in digital design: a standard cell design flow. The goal is to keep the speed and energy performance of the full custom work referenced in Chap. 2. Until now, it lacked the ease of design, the fast design cycle, the predictability of performance and the optimization methodology of modern standard cell design (tool) flows. This chapter solves just that: the industry-standard commercial tool flow is leveraged to provide logic synthesis, timing analysis, place-and-route and detailed power analysis of the logic library presented in Chap. 2. Register-transfer modelling (RTL) can be used to describe functionality. After the logic library is characterized into logic, timing and power models, the logic synthesis can map the RTL to the logic library. A silicon-ready layout with performance verification can be realized during place-and-route.

Section 3.1 digs deeper in automated design strategies and its principles. Section 3.2 sketches the typical VLSI design flow, including logic, physical and library aspects. The targets of the design strategy for this work are discussed in Sect. 3.3. Section 3.4 discusses the proposed design flow at length. The library build-up, characterization and compilation are presented, as well as all the logic and physical implementation aspects typical for the ULV differential systems of this work. Section 3.5 summarizes the implementation details and the effects of the proposed design flow on the proof-of-concept designs presented in Chap. 4. Finally, Sect. 3.6 draws a conclusion.

H. Reyserhove, W. Dehaene, *Efficient Design of Variation-Resilient Ultra-Low Energy Digital Processors*, https://doi.org/10.1007/978-3-030-12485-4_3

3.1 Introduction

Advances in process technologies have enabled single chip circuits of a complexity incomprehensible to a single human being. From the start, design engineers have always found ways to overcome this complexity. *Divide et impera* has been a saying that has surpassed time and it still holds today, especially when considering digital design flows. Key elements in VLSI design and workflow processes are partitioning and hierarchical design. Today, a large set of tools are available to VLSI designers to manage, simulate, generate, test and verify large-scale digital designs. These tools require a significant amount of processing power to manage designs of such a scale. In that way, the evolution in this field could be considered as self-sustaining: complex designs enable more processing power, in its turn delivering more complex designs.

A typical VLSI design flow is organized in multiple levels of abstraction with the goal of hiding details until they become necessary. A possible strategy describing the different levels of abstraction was described earlier in Sect. 1.4. Another approach, dating from the time when VLSI design flows were just being developed, is the Y-chart [2, 9] shown in Fig. 3.1. The Y-chart relates much closer to the actual design flow as it is used today. It considers three domains of representation: functional, structural and geometrical. Every domain has a number of levels, starting at a very high level and eventually descending to individual elements. Again, this concurs with the idea of hierarchical design:

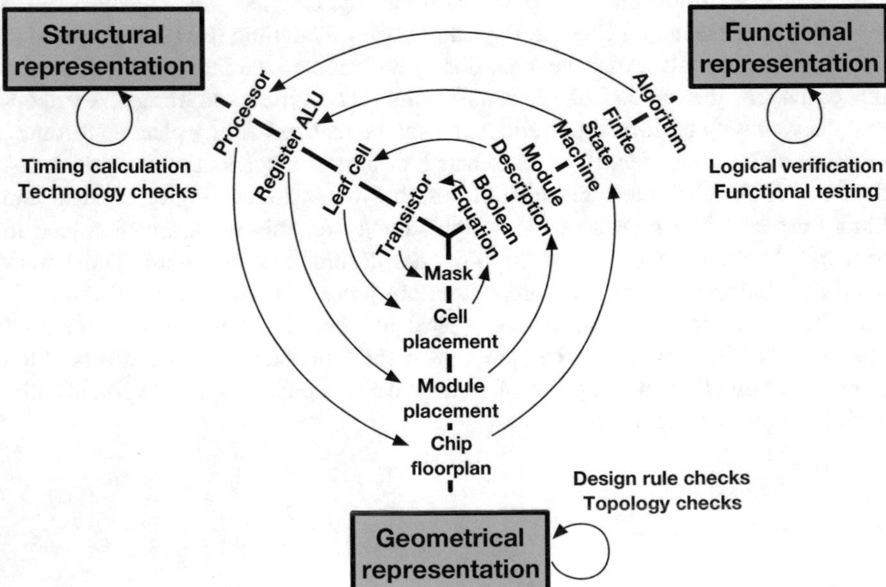

Fig. 3.1 Y-chart showing a common design partitioning strategy [2, 9]

- The **functional or behavioural** branch describes what the system does at the different levels.
- The **structural** representation adds and describes the interconnect between the modular descriptions to achieve a certain behaviour.
- The **physical** representation describes the physical construction of each abstraction level, going from entire device and package to the floor-plan and eventually the transistor geometries and lithography masks.

The Y-chart is of particular interest because its arcs and loops accurately predict or describe the different tools and tasks to come to a final design. Functional loops are typically considered as testing and verification. Structural loops consider timing, power and technological aspects. Geometrical loops include topology and design rule checks. Arcs typically transform one representation into another. High- and low-level synthesis, cell placement and layer extraction are typical examples of such operations.

The standard cell ecosystem with its tools, design flow and abstraction levels is particular to digital design. This origins from the most important level of abstraction in digital design: the use of logic levels instead of voltages and currents. In nanometre scale fabrication technologies, this assumption becomes less and less valid. Parasitic effects, wire resistance, systematic and random performance variations, ambient temperature influence, internal heat dissipation, voltage droop and many other effects can degrade performance to the point where the logic levels are no longer guaranteed. To cope with these effects, standard cell design flows are constantly evolving. EDA tool vendors are now working in close relation with foundries to provide a technology–software combination with a prescribed design flow so that implementations perform as expected. Because managing the design flow (let alone the design itself) has grown so complex, industry has now adopted specialized engineers for almost every part of the design flow. RTL engineers work in close cooperation with characterization, timing, power, verification and test engineers to manage complex designs and create large SoCs. In this regard, the ideal of a standalone automated single-button design flow which leaves no room for error is long gone. The design of billion transistor VLSI designs approaches the edge of what is feasible with today's computing power. A multitude of tools, scripts and managing is necessary to successfully realize state-of-the-art systems. The most recent innovations in standard cell design tools have been on managing the power consumption and the design margin of digital systems. It is no coincidence that exactly these topics are discussed further in this chapter and in Chap. 6.

3.2 VLSI Design Overview

An overview of the typical top-down design flow is shown in Fig. 3.2. It is the best of example of hierarchy and partitioning. The **system and architecture** levels define the functionality and usually form the starting point of the design. Very powerful

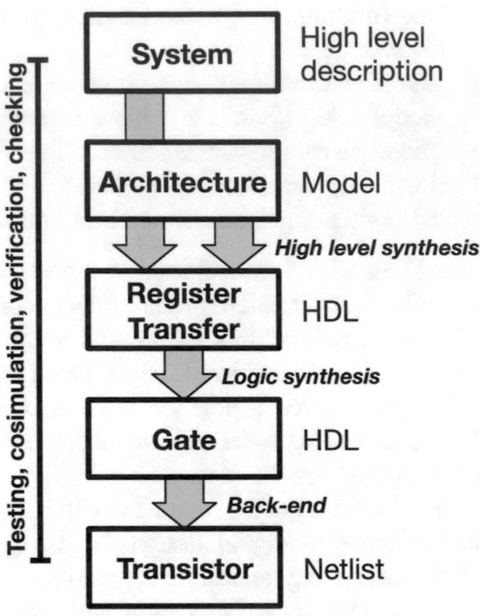

Fig. 3.2 Overview of the typical top-down design flow

tools like Synopsys ASIP Designer allow vertical integration of the architecture design flow. Similarly, high-level synthesis tools are well adopted and can transform C-code models or similar to a **register-transfer description**. The register-transfer model (RTL) describes the synchronous system behaviour in terms of registers and their data flow with its operations. Logic synthesis transforms the RTL to a **gate-level description** consisting of logic gates and flip-flops. Finally, the logic gates are mapped to a circuit-level description consisting of **transistors**, interconnect and other physical cells making up the final chip implementation.

The remainder of this chapter focuses on the bottom three levels: RTL down to transistor description, and the transformation from one to the other. The underlying reason is that the system architecture is usually less aware of technology and power considerations. The RTL, gate- and transistor-level design flows closely interact to achieve the correct functionality, speed performance and power footprint. The outline of the physical design flow is shown in Fig. 3.3. The designer usually doesn't consider the characterization and generation of the standard cell libraries. These steps are discussed in more detail in Sect. 3.2.4. Section 3.2.3 elaborates on the physical design flow.

3.2.1 Physical Architecture

The RTL description closely interacts with the physical design flow to achieve the preconceived specifications. A wide range of architectural techniques are currently being used to get the best system performance possible:

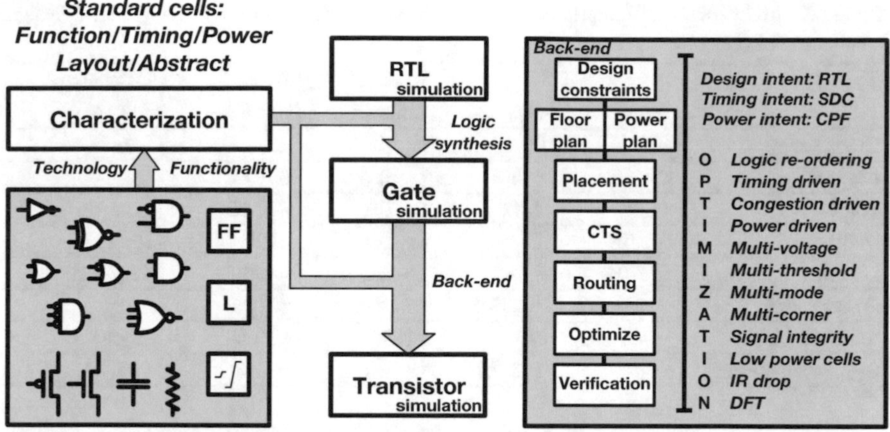

Fig. 3.3 Overview of the physical design flow

- **Clock tree adaptation:** the clock of a synchronous system consumes a significant amount of power due to charging and discharging of the clock network. Partitioning this network, combined with selective gating and correct buffering, can manage the clock power.
- **Logic adaptation:** operand isolation, logic re-ordering and logic resizing can improve the speed performance and reduce the system power consumption.
- **Multi-V_T cells:** standard cell libraries with different threshold voltages swap leakage power for speed performance.
- **Multi-V_{dd}:** partitioning the design into domains with different supply voltages can optimize the power, performance and leakage of those sub-domains.
- **DVFS:** dynamic voltage and frequency scaling controls most of the important factors determining power and speed performance and can thus significantly improve on these.
- **Power gating:** by using power gates, system sub-domains can be powered down when they go unused.
- **Substrate bias control:** providing a bias voltage to the substrate can usefully employ the body effect of the devices to optimize power, performance and leakage.
- **Gate length modulation:** similar to changing the threshold voltage, changing the gate length of individual transistors or cells can swap performance for power.

Table 3.1 gives an overview of a selection of techniques and their impact on the physical design flow, the system performance and the time-to-manufacturing (TTM). The design flow must manage and be aware of all these techniques in an arbitrary complex system without losing oversight and functional results.

Table 3.1 Architectural techniques to improve power with their impact on the physical design, adapted from [4]

	P_{dyn}	P_{stat}	Timing impact	Area impact	Complexity & TTM
Clock gating	20%	0%	~ 0%	<2%	Low
Logic adaptation	<5%	~ 0%	~ 0%	0–10%	None
Multi-V_T	~ 0%	2..3x	~ 0%	−2..+2%	Low
Multi-V_{dd}	40..50%	2x	~ 0%	<10%	High
DVFS	40..70%	2..3x	~ 0%	<10%	Very high
Power gating	~ 0%	10..50%	5..10%	5..15%	High
Substrate bias	~ 0%	10x	10%	<10%	High
L_{gate} modulation	~ 0%	1..2x %	~0%	~ 0%	Low

3.2.2 Logic Synthesis

The first step in physical implementation is logic synthesis. It maps the behavioural description (HDL) to an optimized, technology-dependent, gate-level description. During optimization, the synthesis tool takes into account timing and power information as well as design constraints. Mapping the HDL to logic operations is a CAD field much more related to computer science than to integrated electronics. An extensive overview of the different aspects involved in logic synthesis can be found in [7].

3.2.3 Typical Physical Design Flow

In advanced nanometre technologies, the physical design implementation has become crucial in the development process. If it is performed carelessly, it has detrimental effects on the final system performance. As the diagram in Fig. 3.3 shows, the physical design flow starts with defining the design, timing and power constraints. Next, the floor-plan and power-plan are set out. The cell placement, clock tree synthesis and routing precede post-route optimization. Final checks and verification finish the implementation flow. Without going into further detail, a typical digital design also integrates testability by use of scan chains for post-manufacturing testing. The following steps make up the physical design flow:

1. **Design constraints:** the design constrains define the design targets in terms of functionality (RTL) and timing (SDC). The most recent addition also defines power intent (CPF). Timing constraints include clock period and duty cycle, skew and transition time and input and output delay. Power constraints define power and voltage domains, power nets and 'low-power design' cells such as power switches, isolation cells, level shifters, always-on buffers and retention flip-flops. All design constraints are considered throughout the entire design flow.

2. **Floor- and power-plan:** the floor-plan sets out the allocated chip area, I/O pads, logic domains, macro domains, etc. The power-plan divides the floor-plan into voltage domains and designs the power delivery network down to the standard cells.
3. **Placement:** places the standard cells, macros and low-power cells across the floor-plan.
4. **Clock tree synthesis:** defines the clock delivery network and its specifications. This can be a simple tree-like structure consisting of buffers and (shielded) routing, or a very complex network of clock domains and special routing techniques like the clock mesh discussed in Sect. 4.3.3. Overall, the clock is the most important signal in a synchronous system and can have a large impact on the design.
5. **Routing:** routes the clock tree and interconnects the standard cells, power cells and I/O pads.
6. **Post-route optimization:** performs optimizations often only available after the previous steps have been completed. Leakage recovery swaps high leakage cells with lower leakage cells on non-timing-critical paths to save leakage power. Hold time optimization buffers short paths to guarantee correct functionality and is only executed in the final design phase to prevent overly conservative hold time fixing.
7. **Verification and sign-off:** is a final step functionality and performance is verified with the most detailed technology data before going to design fabrication.

Apart from the typical sequence of physical implementation steps, things like IP integration, timing and power analysis, multi-mode multi-corner optimization, accounting for on-chip variation and design margins are considerations in the physical implementation flow.

3.2.3.1 IP Integration

Digital designs today often combine logic functionality mapped to standard cells with soft or hard macros of acquired intellectual property (IP). These macros usually come with dedicated timing libraries and layout views. It is not unusual that a large part of the physical design flow consists of combining these IP blocks with design-specific functionality to create a system-on-chip.

3.2.3.2 Static Timing Analysis

Since the most important design target has always been speed performance, every step in the physical design flow is timing-aware. Floor-planning, placement, CTS, routing and optimization all include delay calculations in the form of timing analysis to check the system speed performance. Whether or not the design achieves that speed performance is the main consideration in the decision-making process of all the optimizations involved. The term *static* is used to describe the fact that the delay calculation is not data-aware. It simply adds the delay of the different logic gates that contribute to a single logical path between two registers and compares it to the available propagation time, taking into account the clock. Depending on the abstraction level, influences like interconnect RC delay and signal crosstalk are accounted for in coarse or fine detail.

3.2.3.3 Power Analysis

While the power consumption is not a quantified design target, power analysis and power-aware optimization are equally important as timing. In large-scale designs, simply meeting timing specifications without considering power can result in infeasible constraints on the power delivery network and the heat dissipation which again impacts timing. Static and dynamic IR drop analysis help to account for these effects.

3.2.3.4 Multi-Mode Multi-Corner (MMMC)

Advanced nanometre CMOS technologies have significant performance differences depending on process corners, temperature and supply voltage. Moreover, the architectural techniques presented in the beginning of this section give a system various operating, voltage and frequency modes. The combination of all these influences results in an exponentially increasing collection of operating points that need to be addressed during timing analysis, power analysis and physical implementation (see Fig. 3.4). Optimization techniques are aware of all these modes and can concurrently optimize the design with different constraints for each mode: multi-mode multi-corner (MMMC) optimization. The goal is to optimize the overall performance, rather than favouring one operating mode for the other. It is clear that the required computational resources necessary for co-optimization of an increasing

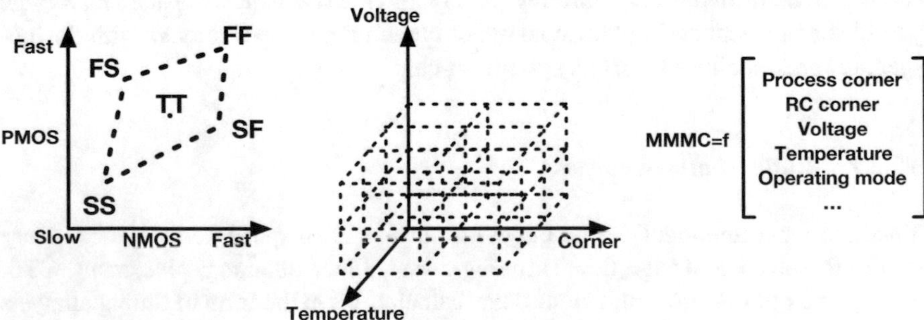

Fig. 3.4 Multi-mode multi-corner optimization concept

number of modes and corners increase super-linearly. Often, a subset of modes and corners is chosen for optimization, while the full set is verified during sign-off.

3.2.3.5 On-Chip Variation (OCV)

Until recently, standard cell libraries did not account for intra-die process variations of the on-chip devices. Most models only account for global variations and provide different library files for different process conditions. In reality, gates exhibit variations in delay and power due to both inter- and intra-die process variations. To account for this, physical implementation flows *derate* timing conditions, in fact scaling the delay with a user-defined factor. For setup time analysis, clock paths are sped up (derating factor <1), while logic paths are slowed down (derating factor >1). For hold time analysis, clock paths are slowed down (derating factor >1), while logic paths are sped up (derating factor <1). Accurately determining the necessary derating factors to guarantee correct performance without overdesign is non-trivial and in fact requires detailed knowledge of the standard cell performance under variations. In practice, standard cell libraries often come with their dedicated OCV derating factors.

3.2.3.6 Design Margins

The use of MMMC and OCV requires caution. The risk of conservatively stacking design margins is high. Combining the right operating modes with their own constraints and OCV parameters is an important part of the design that requires input from all the abstraction levels of the design flow. While most of the techniques discussed in this section aim to reduce power, over-design typically results in a power overhead. Eliminating design margins and overcoming their power overhead is discussed extensively in this work in Chap. 5 and 6.

3.2.3.7 Recent Developments

Recent developments in the physical design flow aim to reduce design margins and predict system performance more accurately. **Statistical static timing analysis (SSTA)** is a promising technique that aims to overcome pessimistic delay calculation while taking into account on-chip variation. In this case, the library provides mean and standard deviation of the delay, in order to more accurately estimate full path delays, as shown in Fig. 3.5. Similar to MMMC, the approach requires a lot of computational resources, since every path delay calculation now requires a lot more computations.

Stage-based on-chip variation estimation (SBOCV) is another promising approach that keeps middle ground between SSTA and OCV. Stage-based OCV exploits path length-dependent OCV, based on the idea that long paths are more

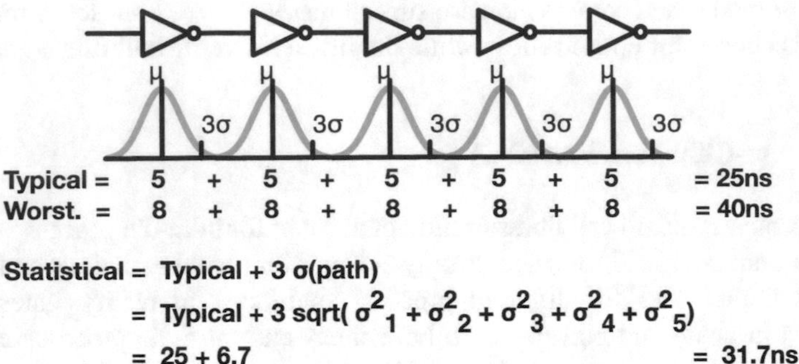

Fig. 3.5 Concept of statistical static timing analysis

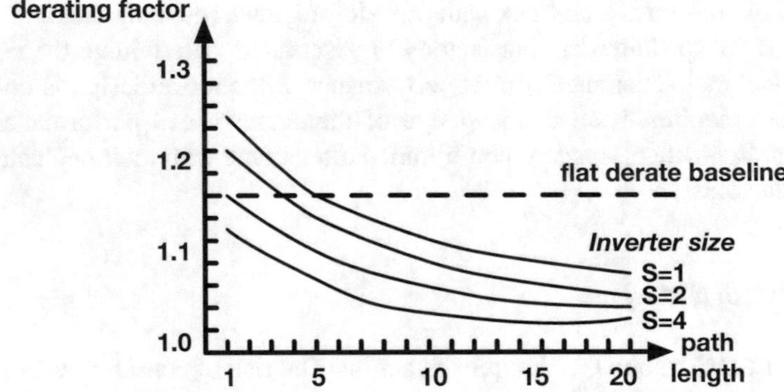

Fig. 3.6 Concept of stage-based OCV, employing an OCV derating factor according to path length

likely to level out on-chip variations because not all cells are simultaneously fast or slow. In this case, the library provides different timing deratings according to the cell function and size and the length of the path-under-test, as can be seen in Fig. 3.6. SBOCV could be used standalone or as a fast approximation to statistical static timing analysis. Both approaches target sub-20-nm CMOS technologies, but could be promising for variation-sensitive conditions such as ultra-low-voltage operation of larger feature size technologies.

3.2.3.8 Tools

The physical design flow described in this section is largely tool-driven. EDA vendors such as Cadence, Synopsys and Mentor Graphics are the main drivers of these and work in close cooperation with silicon fabs to fine-tune the tools to the technology. Because of the complexity of today's digital designs and technologies, designers have to rely on tools and algorithms to do most of the work. Although these extensive tool suits are very powerful, the result still depends on how well the designer can express his or her fundamental design ideas to guide the design to the desired outcome.

3.2.4 *Typical Standard Cell Library*

Section 3.2.3 omitted the details on standard cell libraries and how they are conceived. This section sketches a typical standard cell library and how it is characterized into workable models. A standard cell library is a collection of logic cells made up of primitive or macro functions that form the basis for a digital design. The main goals are predictability, ease of design, error avoidance and re-usability. In light of the divide-and-conquer approach of VLSI design, standard cells can be considered as black boxes limited to the behaviour as modelled by the library files. Next, the cell details, library details and characterization details are discussed.

3.2.4.1 Standard Cell

Figure 3.7 depicts six representations of the same standard cell, each providing its own information. The considered cell is a simple inverter, but the same representations hold for complex logic gates, flip-flops, etc. These six representations provide a complete model of the standard cell and are enough to implement an SoC.

- **The logic view** provides the functionality of the cell, essential for logic synthesis and optimization. Logic views can be simple (inverter, 2-input and, ...) or much more complex (flip-flop, and-or-invert, ...).
- **The netlist** sketches the transistor implementation including sizing of the different devices. In some cases, a parasitic extraction of the layout is also included in the netlist to allow more accurate modelling.
- **The timing view** describes input-to-output delay for the different data transitions, as well as rise and fall times. A table provides delays for a range of input slew rates and output load capacitances. As conditions are rarely exactly as described in the table, intermediate values are interpolated. For more sophisticated cells, other timing relations are described in a similar manner. A flip-flop also describes metrics such as setup time, hold time, minimum clock width, removal time and recovery time.
- **The power view** considers the same events as the timing view and provides the power consumption in each case, as well as the (data-dependent) leakage power.
- **The layout view** provides the detailed layout geometries at the transistor level. Typically, this view is too detailed for the physical implementation flow, so it is only used to extract the abstract view and during final sign-off checks and tape-out. To facilitate physical implementation, standard cells are designed according to a fixed height, or a multiple thereof. This allows placement in rows. Metal interconnect wires are placed on a grid according to the technology design rules.

Fig. 3.7 Full standard cell description of a simple inverter cell

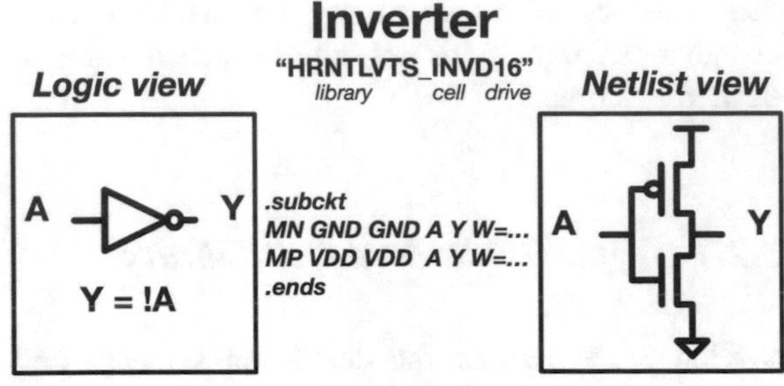

Timing view

!A→Y (input slew \ output load)			
input slew	0.2ns	11.4ns	90.4ns
	1.1ns	12.7ns	91.6ns
	3.6ns	20.7ns	99.7ns
	5.4ns	28.3ns	107.9ns

A→!Y (input slew \ output load)			
input slew	0.2ns	10.2ns	89.1ns
	1.0ns	11.7ns	90.6ns
	3.3ns	19.7ns	97.7ns
	5.0ns	26.3ns	105.9ns

Y↑ (input slew \ output load)			
input slew	0.09ns	1.58ns	6.88ns
	0.37ns	1.72ns	6.95ns
	0.84ns	2.40ns	7.11ns
	1.44ns	3.69ns	7.94ns

Y↓ (input slew \ output load)			
input slew	0.08ns	1.54ns	6.37ns
	0.35ns	1.56ns	6.75ns
	0.80ns	2.20ns	7.01ns
	1.31ns	3.51ns	7.30ns

Power view

When	Leakage power
A	1.5nW
!A	1.7nW

Y↑ (input slew \ output load)			
input slew	0.26nW	1.4nW	34nW
	0.45nW	1.5nW	35nW
	0.86nW	4.64nW	52nW
	1.55nW	9.8nW	71nW

Y↓ (input slew \ output load)			
input slew	-0.01nW	-0.15nW	-0.35nW
	-0.04nW	-0.17nW	-0.35nW
	-0.07nW	-0.26nW	-0.36nW
	-0.09nW	-0.31nW	-0.36nW

Layout view

VDD
Y
A
GND
fixed height
variable width

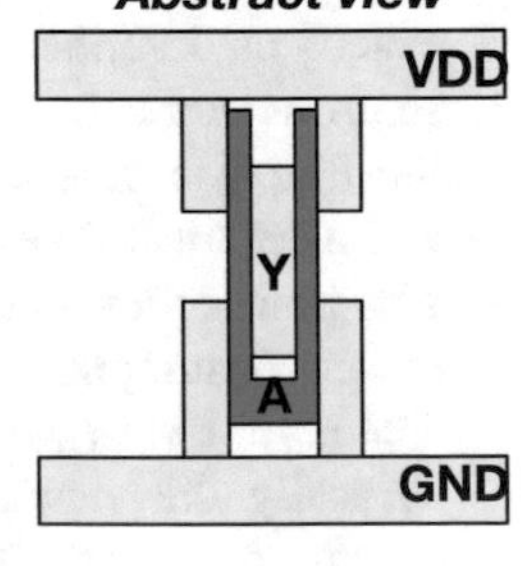

Power and ground pins as well as bulk connections are located on the top and the bottom. These concepts allow standardization of the cell interfacing.
- **The abstract view** is derived from the layout to provide a simple cell representation for the physical implementation flow. It consists of metal wires, cell access pins and metal blockage.

Most of this information is acquired through characterization of the cell and is summarized in the library model files.

3.2.4.2 Library

A standard cell library gathers a collection of standard cells and its different representations. A variety of logic functions and sequentials is always available, as well as different drive strengths. Ideally, the logic synthesis tool and optimizer should have enough freedom to trade off between delay, area and (leakage) power. Complex cells such as half adders, full adders, and-or-invert or similar improve the efficiency of the logic mapping. Implementation with different V_T devices typically allows a good delay and leakage power optimization. A typical standard cell library has the following functionality:

- Simple logic functions such as INV, NAND, NOR, XOR, XNOR, MUX, etc., all with different drive strengths and V_T devices.
- Complex logic functions like half adder, full adder, AOI, OAI, MAOI etc., all with different drive strengths and V_T devices.
- Sequential elements like flip flops or latches, positive and negative edge or level triggered with different drive strengths.
- Special inverters or buffers with balanced rise and fall delays typically used for clock signal buffering.
- Delay cells necessary for short path padding to fix hold time violations.
- Low-power cells such as level shifters, isolation cells, state retention flip-flops, power switches, always-on buffers, etc. to accommodate low-power physical implementation.
- Scan flip-flops to enable scan chain insertion for testing purposes.

For physical implementation, the full library is summarized in three types of representation: functional (Verilog or similar), timing/power (liberty) and physical (LEF). These give the physical design flow enough information to create a reliable design. The *Verilog* file provides the logical function of the cell through a truth table or a logic function. The *liberty* file contains all the timing and power information, input capacitances, output loads and slew rates. These are enough to model cell behaviour during synthesis and place-and-route. Finally, the *LEF (library exchange format)* file provides the abstract views of each cell so placement and routing happen in accordance with the physical layout of each cell.

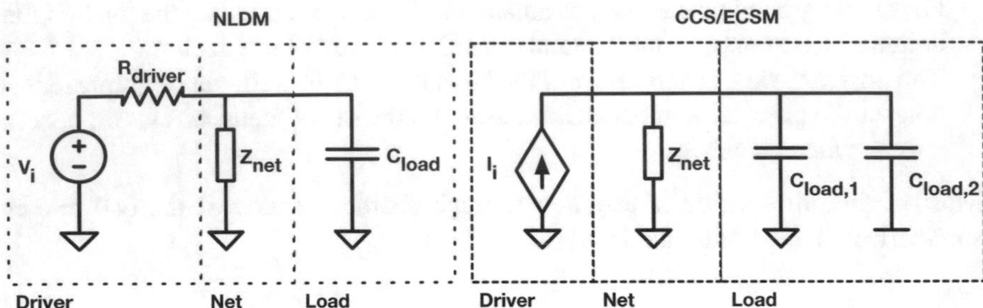

Fig. 3.8 Characterization approach in NLDM and CCS/ECSM models

3.2.4.3 Characterization

Characterization aims to make a sparse model of the behaviour and properties of each individual cell. The use of a good-enough modelling approach is what makes standard cell design so powerful. A lot of the circuit details are omitted, thus enabling faster design and analysis. The re-use of standard cell libraries in multiple designs usually justifies the fact that they require a lot of effort to generate. The biggest part of characterization consists of taking the technology SPICE models and simulating the circuit-under-test in a variety of conditions to correctly capture its behaviour. Modern characterization tools automate most of the process. They sense cell functionality, run simulations under realistic conditions, extract the simulation results and compile the libraries. Throughout this work, Synopsys SiliconSmart was used for cell characterization.

Different modelling approaches are used in characterization. The most popular is NLDM (non-linear delay model). CCS (composite current source) and ECSM (effective current source model) are being used more recently. All three approaches use a driver-net-receiver modelling technique, as shown in Fig. 3.8. NLDM uses a voltage source with the appropriate impedances and a simple load model. CCS uses a current source with net impedance and a dual load model that uses different output loads depending on the point in the delay curve. ECSM also uses a current source as the driver, but captures the voltage waveform at the output load instead of the current through the load (CCS). NLDM captures only three output points: delay threshold and transition threshold (typically 50%, 10% and 90% of the output voltage). The main benefit of current source modelling over voltage source modelling is its inaccuracy when $R_{driver} << Z_{net}$, and the more complex load model able to represent effects like the Miller capacitance.

3.3 Goal

This chapter aims to facilitate efficient design using the logic gates described in Chap. 2. The design flow described in the previous sections is unrivalled when it comes to efficiency, predictability and design time. The standardized tools and algorithms are industry-proven. We target their use as much as possible, as well as the standard cell design principles. The final goal is a design flow that leverages the good properties of the transmission gate-based full custom designs, with all the nice properties of tool-based VLSI design. More specifically, the following aspects are desirable:

- Create differential transmission gate logic **standard cells**, fully characterized with timing and power information.
- Use the (physical) design flow with commercial tools (as much as possible) to facilitate an **easy and fast** design cycle.
- Take **ULV operation** in consideration during the entire flow contemplating overdesign and design margins.
- Provide performance **predictability** in terms of speed and power.
- Gain insight in **minimum energy** consumption through multi-V_{dd} simulations.
- Design **large** digital systems of a scale infeasible to the full custom design approach.

All these goals are achieved in the design flow proposed in Sect. 3.4. More practically, the differential transmission gate logic of Chap. 2 is not plug-and-play compatible with the typical standard cell flow. Differential logic synthesis is not available in commercial tools. The proposed tool flow provides a work around by enabling single-ended synthesis and post-synthesis transformation to a differential gate-level netlist.

3.4 Proposed Design Flow Approach

An overview of the proposed design flow that achieves the goals of Sect. 3.3 is depicted in Fig. 3.9. The flow chart describes the different steps to implement a full low-power digital system (e.g. the microcontrollers of Chap. 4) using differential transmission gate logic. Commercial EDA tools are used as much as possible to realize each operation. The additions to the flow developed in this work are shown in grey. They are typically realized through scripts in *TCL* or *Python* and intervene in the normal standard cell flow when possible. The full flow chart can be split up into three chronological levels:

1. **Library level:** defines, sets up and characterizes the differential transmission gate library. Moreover, it transforms the differential library to a pseudo single-ended library, containing the differential timing/power/area information.

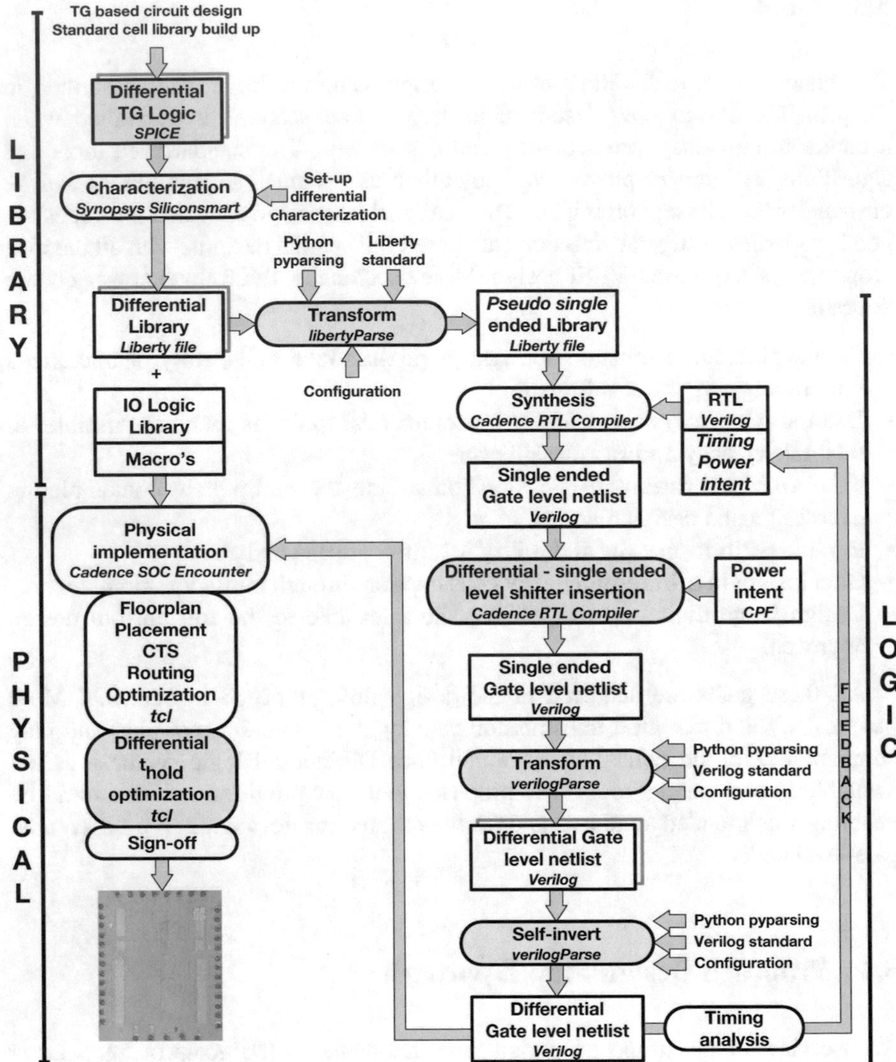

Fig. 3.9 Proposed differential transmission gate design flow chart. Grey boxes are augmentations to the typical flow using custom scripts

Transformation is realized through a custom liberty file parser and manipulation tool.

2. **Logic level:** synthesizes the RTL system description into a single gate-level netlist using the pseudo single-ended library. The single-ended gate netlist is transformed to a differential gate netlist using a custom verilog file parser and module/gate/pin/net manipulations. Differential to single-ended domain interfacing is realized through level shifters using the low-power implementation

flow. Finally, the proposed *self-invert technique* removes all inverters from the design and replaces them with their differential signal counterparts. This slims the gate-level netlist and improves power consumption. Signal buffering to improve performance is done during physical implementation.

3. **Physical level:** implement the physical system using the differential netlist and differential library. The physical implementation is timing and power aware and has the actual information from the differential cells. It goes through the different implementation steps just like any other (single-ended) design flow. Finally, for hold time optimization some intervention in the differential signals is required.

3.4.1 Library Level

3.4.1.1 Cells

The standard cell library is mostly made up of the logic functions and flip-flops. Transmission gates and inverters make up nearly all cells. By cascading multiple transmission gates, complex functionality can easily be achieved (see Sect. 2.2.2). All standard cells are constructed according to the diagram in Fig. 3.10. The most simple logic cells are 1-level transmission gates and provide elementary 2-input logic functions. More complex 3- and 4-input logic cells can be created using 2-level transmission gates. The most complex cells use 3-level transmission gates and provide 5- to 8-input logic functionality. Figure 3.11 shows some examples of how these cells are built according to the scheme in Fig. 3.10. For 2- and 4-input logic functions, it makes sense to create cells with every logic function. For 5- to 8-input logic functions, only AND, OR, NAND and NOR operations are implemented. Every cell is created in three drive strength variations (1, 2 and 4) according to the

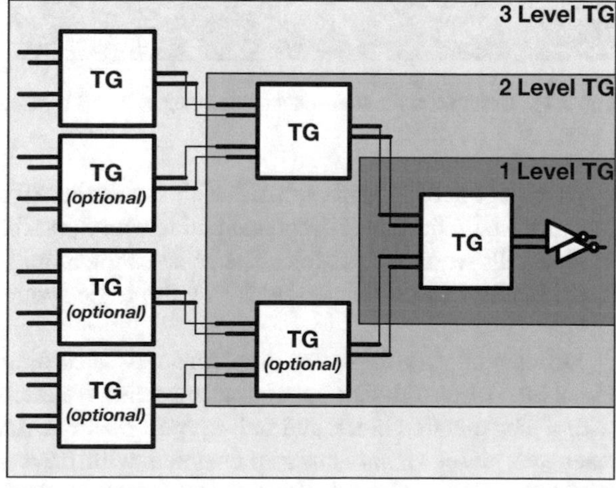

Fig. 3.10 Standard cell deconstruction into transmission gate and inverters

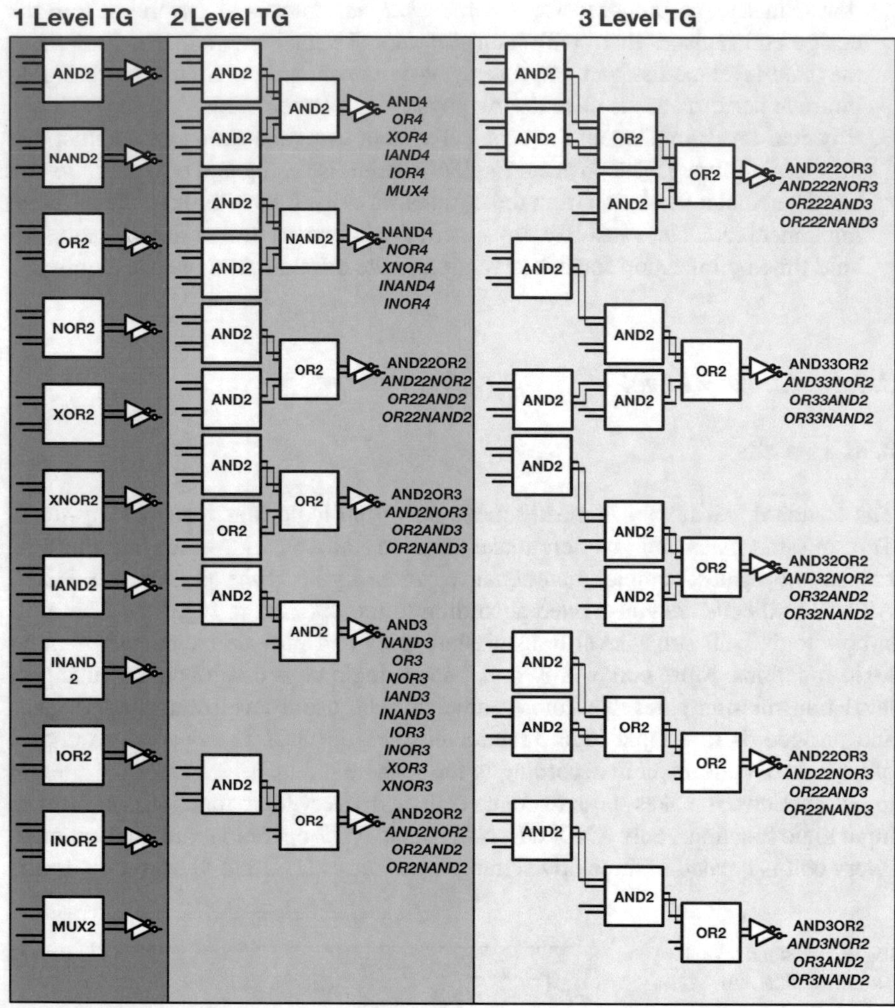

Fig. 3.11 Example logic implementations using 1, 2 or 3 levels of transmission gates

inverter stage at the end of each cell. This results in 189 cells. Note that if every logic function would have been implemented in every possible combination, more than a million cells would be feasible. The subset shown in Fig. 3.11 is chosen because its functionality was considered useful by the logic synthesis tool during preliminary tests.

Differential flip-flops are implemented according to the circuit shown in Fig. 2.30. A transmission gate-based master/slave latch structure is used, together with a single-ended clock and pull-up/pull-down devices for reset. Flip-flops with reset and preset signals are implemented with drive strength 1, 2 and 4. In the implementation in Chap. 6, timing error detection flip-flops are added as well.

Apart from logic functions and flip-flops, a number of additional cells are implemented in the standard cell library:

- Clock inverters with balanced rise/fall time and faster propagation delay, implemented with drive strength 1, 2, 4, 8, 16, 32, 64, 128, 256 and 512.
- Level shifters providing interfacing between both differential to single-ended domains and ultra-low-voltage to I/O-voltage domains. In this work, the ultra-low-voltage domain is always implemented differentially, while the I/O-voltage domain is single ended. A unified level shifter for each domain crossing thus provides both functionalities. This is quite convenient, since level shifting from a low to a high voltage usually requires differential input signals to control their differential amplifier structure.
- A clock level shifter is implemented apart from the normal level shifters to facilitate low-delay level shifting at a higher power/area cost.
- Hold time buffers are implemented separately using a combination of inverters and transmission gates as intermediate loads to provide the best delay/area/power trade-off.
- Filler cells to fill out standard cell gaps in the final implementation have no logic functionality, so they can easily be added to the library.
- I/O-voltage cells with simple functionality such as INV, BUF and NAND2 are implemented to enable full operability in the I/O domain.
- I/O pad cells are implemented to complete the physical design flow. Power pads, signal input pads, signal output pads, clock pads and input/output pads are implemented and characterized.

3.4.1.2 Characterization

As shown in Fig. 3.9, the library is characterized using Synopsys Siliconsmart. The differential cells require a slightly different approach. Normally, the tool characterizes each cell under every possible permutation of input signals. For differential cells, the condition that two differential inputs should always behave differentially is added. For a differential AND2 gate, this results in the following logic description:

```
input A, A_BAR, B, B_BAR
output Y, Y_BAR

Y = (A&!A_BAR) & (B&!B_BAR)
Y_BAR = !((A&!A_BAR) & (B&!B_BAR))

switching set {A A_BAR}
switching set {B B_BAR}

forbidden state { !(A^A_BAR) }
forbidden state { !(B^B_BAR) }
```

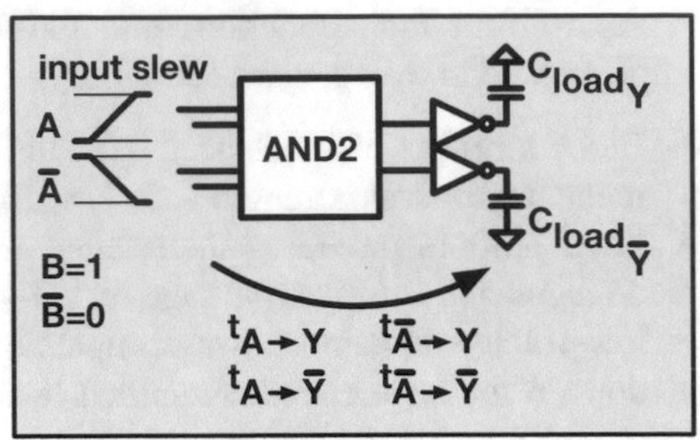

Fig. 3.12 Example characterization setup of a differential AND2 cell

If the differential signals would not switch together, undefined logic states would arise. The propagation delay and output load of differential signals is modelled individually. Figure 3.12 demonstrates this concept. Timing arcs from each input pin to each output pin are in fact generated. This enables separate pin analysis during static timing analysis and does result in a separate delay and power for each individual input pin. Static timing analysis then takes into account differences in slew rate, propagation delay and output load and thus provides reliable information on the differential operation. More complex cells follow the same principles. All cells are characterized using NLDM models. Characterization of differential cells requires significantly more resources than single-ended cells. Especially when multiple low voltages and process corners are applied in combination with complex gate functionality, characterization of a full library can take weeks on a decent number-crunch system.

3.4.1.3 Differential to Single-Ended Transformation

Using differential cell models is not possible in commercial synthesis tools. To overcome this challenge, the proposed design flow transforms the just characterized differential standard cell library into a *pseudo* single-ended standard cell library.

Compiling a single-ended library out of a differential library comes down to merging the differential function, pins and characterized information. Different merging strategies are possible. In this work, a *worst case merging* approach is always used (see Fig. 3.13). In practice, this means that a differential cell maps a pair of differential pins to a single-ended pin by taking the worst of the two. This is done for timing properties, power properties and capacitances. A necessary condition for merging two differential pins is that they are characterized under the same slew and load conditions. This is reasonable, as differential pins often exhibit similar conditions and come from the same device structure.

The library transformation is automated using a custom script set. The open-source liberty file format standard [8] is used as a basis to create a liberty file parser in Python. The Python Pyparsing package [3] helps to define the liberty syntax in an intuitive way. The library is read in the database and mapped to an object-oriented model of the liberty syntax. The cell objects are then manipulated with custom functions like *mergeData(rise_transition)*, *mergeTiming(pin* A, *pin* $\overline{\text{A}}$),

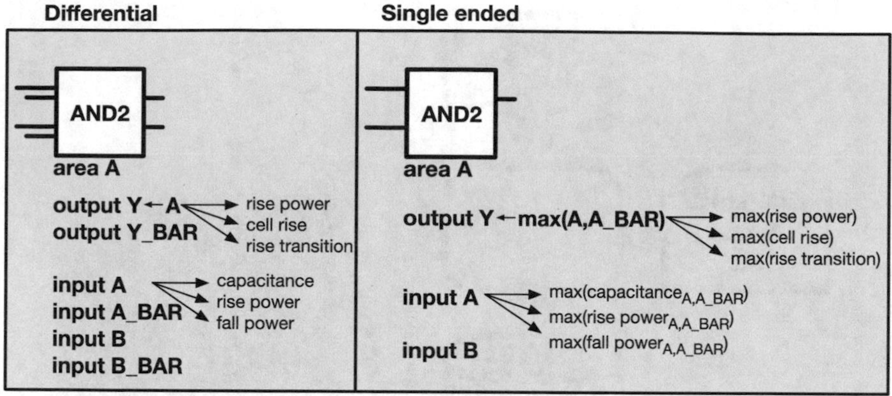

Fig. 3.13 Differential to single-ended mapping of characterized data, applied on an AND2 gate. A worst case merging approach is applied

mergePin(pin A, *pin* $\overline{\mathrm{A}}$) and *merge(pin* Y, *pin* $\overline{\mathrm{Y}}$). Finally, the single-ended library is printed to a text file. The data flow of the library manipulation is shown in Fig. 3.14.

3.4.1.4 *Pseudo* Single-Ended Library

The transformation from differential to single-ended domain results in a *pseudo* single-ended standard cell library. *Pseudo* referring to the fact that it does not have any physical meaning, other than having functionality, timing, power and area information related to the actual differential library. The library does not have a functional layout, so physical implementation is not possible. However, before physical implementation can happen, the RTL functionality has to be mapped to physical gates, which occurs at the logic level of the proposed design flow.

3.4.2 *Logic Level*

3.4.2.1 Synthesis

Logic synthesis is a core functionality of VLSI design, since it performs one of the crucial steps in descending levels of abstraction. The power of logic synthesis is that it incorporates the timing and power information of the basic logic functionality of the standard cells and uses them as cost functions in the logic mapping and optimization. Logic synthesis EDA tools incorporate a collection of algorithms and mapping techniques, combined with other useful functions. They are so well tailored to today's advanced technologies that VLSI design cannot but employ them. The

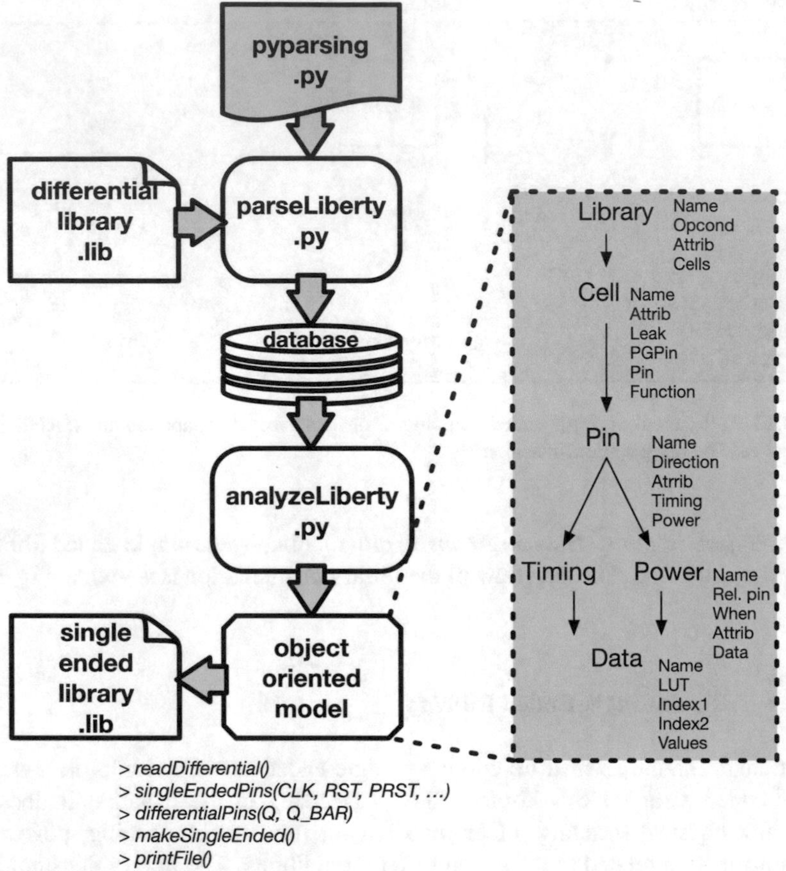

Fig. 3.14 Standard cell library single-ended transformation flow

proposed design flow uses them as well, leveraging their power for low-power and efficient design. Because logic synthesis uses the single-ended libraries as described in Sect. 3.4.1, there is nothing that limits the full functionality of the synthesis tools in the proposed design flow.

Transmission gate logic implements non-inverting and inverting logic with the same building block. The logic synthesis can thus use non-inverting functions without an additional delay or area penalty. This is clearly shown in the microcontroller implementations discussed in Chap. 4. Figure 3.15 shows synthesis results from both prototypes. The pie charts show that a significant percentage of logic gates is mapped to complex logic functions (2 and 3 level transmission gates). The bar diagrams show the most commonly used logic functions. The synthesis tool clearly maps large parts of the system to non-inverting logic.

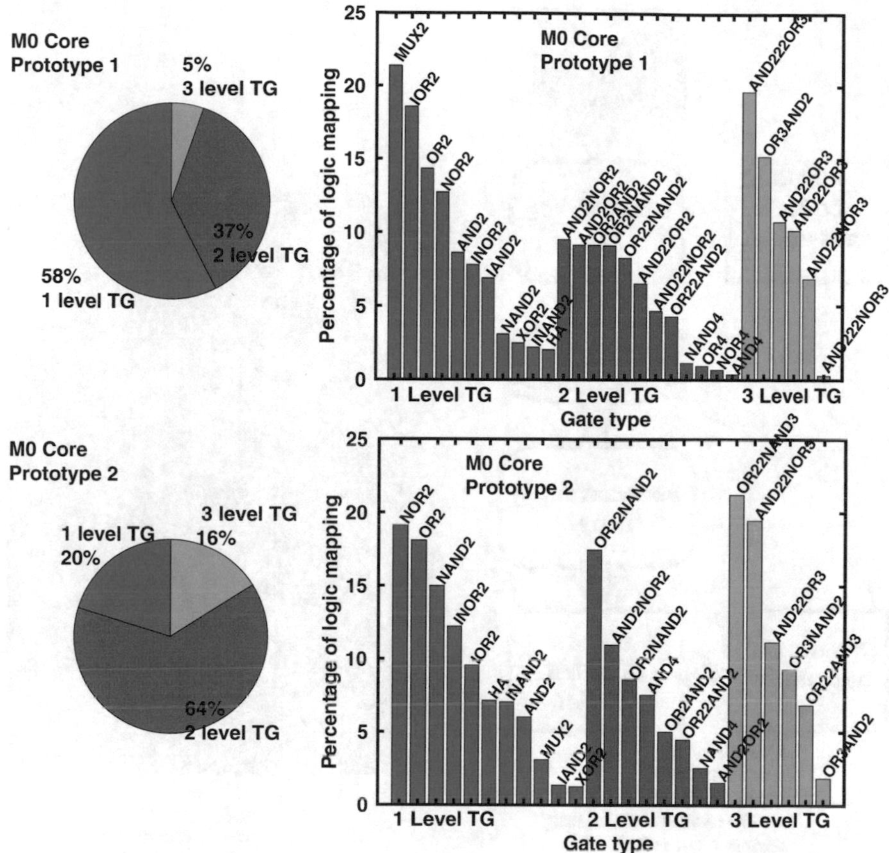

Fig. 3.15 Logic function usage for the microcontroller systems of Chap. 4. Complex multi-input gates and non-inverting functions are used extensively

3.4.2.2 Single Ended to Differential

The single-ended synthesis results in a single-ended gate-level netlist, in accordance with the *pseudo* single-ended standard cell library. As much as this netlist relates to the library used for synthesis, the goal is to implement a differential system. A crucial step at the logic level is thus to restore the differential gate topology and its accompanying standard cell library in the gate-level netlist.

Compiling a differential gate-level netlist from a single-ended gate-level netlist comes down to mapping single-ended modules/cells/functionality/pins to their differential counterparts, in accordance with the original differential library. Special consideration is necessary when converting some cells: flip-flops have single-ended functionality attached to clock and reset pins, while differential functionality is attached to their output (Q and $\overline{Q}$); SRAM macros remain in the single-ended I/O

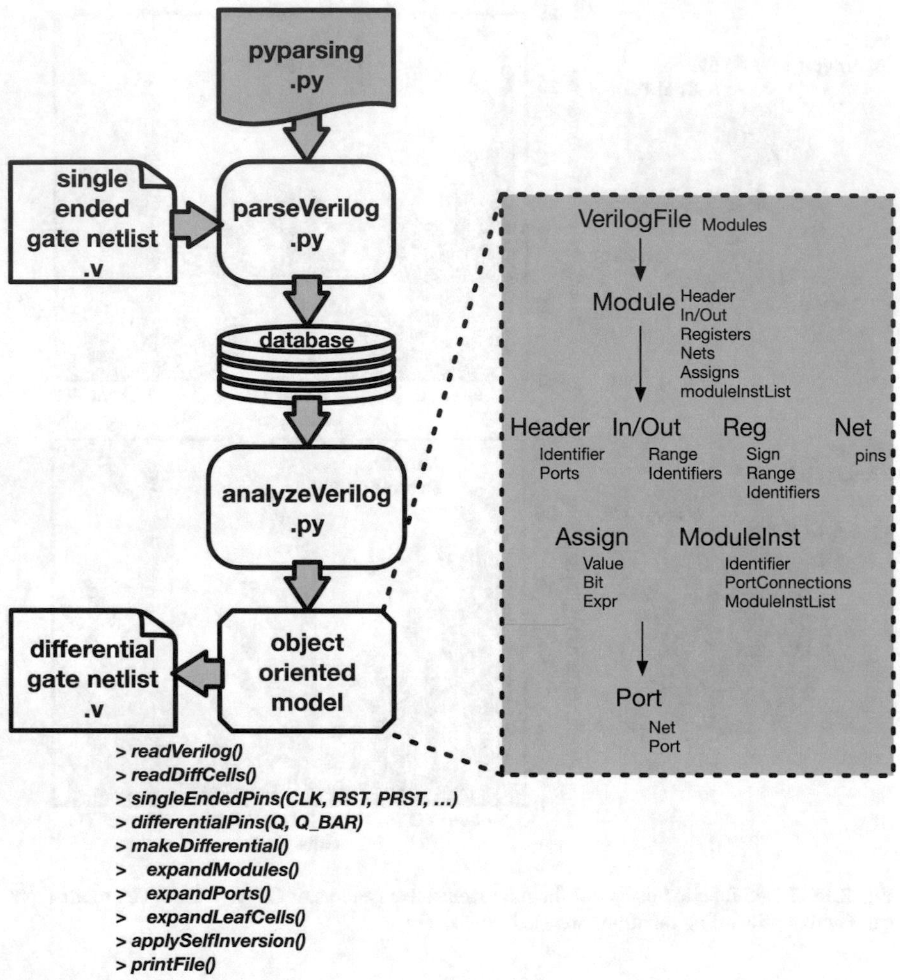

Fig. 3.16 Gate-level netlist differential transformation flow

domain and require the correct interfacing, as do other I/O domain signals and building blocks.

The netlist transformation is automated using a set of custom scripts. Figure 3.16 shows the transformation flow in more detail. Similar to the library transformation in Sect. 3.4.1, the Pyparsing package [3] is used to read any Verilog gate-level netlist in the database and map it to an object-oriented structure. A range of manipulations can then convert the netlist to a differential netlist and write it to a new file.

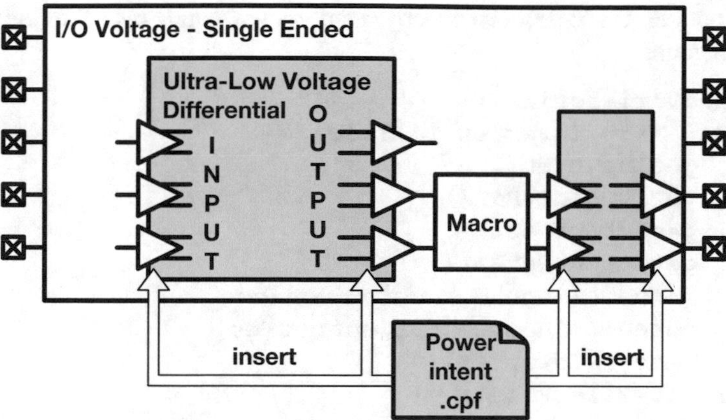

Fig. 3.17 Cross domain interfacing between differential and single-ended domain occurs through power intent specification

3.4.2.3 Cross Domain Interfacing

The differential transmission gate library can be used in conjunction with a single-ended library, as shown in Fig. 3.17. This is the case when the system also uses single-ended logic (for example, at a higher V_{dd}) or employs single-ended macros. In this case, the low-power flow is leveraged to limit the differential domain to a specific power domain. The power intent file designates level shifters that handle the differential to single-ended domain crossing and vice versa. In the proposed flow, these level shifters immediately take care of voltage level shifting as well.

Although level shifting could be considered as something specific to the physical implementation, in the proposed flow they are inserted at the logic level. Differential to single-ended domain switching is much more a logic procedure, and functional simulations would not be possible without these level shifters in place. By defining the cross domain interfacing in the power intent file, any optimizations or operation afterwards are aware of the domain crossing, and incorporate the level shifters.

3.4.2.4 Self-Inverting Technique

An inherent advantage of using differential logic is that complementary signals are readily available. Looking at the logic operations, having inverters in the netlist does not make sense. Especially during signal buffering, clock tree synthesis and short path padding are only performed during or after physical implementation. All the inverting operations can be realized by connecting the complementary output to the regular input and vice versa. In the current object-oriented representation of the Verilog netlist, these operations are easily accomplished. The following example code describes the operations incorporated in the single-ended to differential transformation:

```
> readVerilogFile()
> for every Inverter in File:
>     getInputNet()
>     getInputNetBar()
>     getOutputNet()
>     getOutputNetBar()
>     connect(inputNet,outputNetBar)
>     connect(inputNetBar,outputNet)
>     remove(Inverter)
> printVerilogFile()
```

The prototypes presented in Chap. 4 demonstrate the benefits of this technique. Prototype 1 does not use the just discussed self-inverting technique, while prototype 2 does. For the M0 system, the self-inverting technique removed 3894 inverters, which was close to 35% of the logic gates. The net area overhead of these (differential) inverters was 46521 μm^2. Note that the final area impact is hard to evaluate since it is also significantly influenced by other physical implementation effects like placement density, design-rule-check buffering and logic optimizations.

3.4.2.5 Feedback

Because the logic synthesis tool relies on the single-ended library rather than the differential library to optimize the design, the timing and power intent of the synthesis is no longer guaranteed for the differential netlist. Because the *pseudo* single-ended library is composed of worst case-based data, this is effect is minimized. A detailed timing analysis of the differential gate-level netlist using the differential library can indicate whether the differential netlist can meet the intended timing constraints. If not, this information should be used to iterate the logic synthesis process and adjust the initial timing constraints. While the proposed synthesis flow may be slightly slower due to single-ended to differential transformation, the automated process greatly simplifies this. Moreover, logic synthesis is, by far, the most time-consuming operation in this process.

To make sure the just described iteration converges, timing constraints are preferably set tighter than intended in the final physical implementation. This design practice is already well adopted in advanced technology physical implementation flows where design constraints are relaxed after each step of the process. For the microcontroller systems of Chap. 4, differential timing analysis after single-ended synthesis (but before physical implementation) required a 15% tighter constraint than targeted under slow-slow process conditions.

3.4.3 Physical Level

Physical implementation is the final step in the proposed VLSI design flow. Commercial EDA tools have grown to the extent that they incorporate logic functionality, timing and power during each step and optimize accordingly. The proposed design flow leverages this functionality as much as possible in a semi-automated design process. Essential steps include the layout of each cell, placement, routing, clock tree synthesis, leakage optimization and hold time optimization.

3.4.3.1 Cell Layout

The differential logic library consists of transmission gates and inverters only. This property is cleverly used to layout the full digital library in an automatic way, as represented in Fig. 3.18. The TG and inverter building blocks are constructed in SKILL code [1]. Next, logic functions, device gate length and interconnect options are added depending on the desired logic function of the cell. This strategy is quite efficient and results in a very dense layout of every cell, partly thanks to the limited amount of possible interconnect options.

Flip-flops, level shifters and clock buffers are much less regular. The layout of these building blocks is custom made to achieve the best cell density.

3.4.3.2 Floor-Plan, Placement, CTS and Routing

As described in Sect. 3.2.4, the abstract views of the layout are used to continue physical implementation. Designing the **floor-plan**, **placing** the cells, implementing a **clock tree** and routing the interconnections all happens very much in line with the typical physical design flow as described in Sect. 3.2.3. Particularities like power-aware floor-planning and mesh-based CTS are much more related to the ultra-low-voltage operation of the system than to the differential properties of the design and are discussed in detail in Chap. 4.

Routing is expected to be more prone to congestion due to the differential nature of the logic. After all, differential signalling by approximation doubles the amount of signal interconnections. Congestion is also influenced by other factors including the timing constraint, the amount of routing tracks, the standard cell height, the technology metal stack, the core density and the power distribution network. In the physical implementation flow, all these factors closely interact, making it hard to see the influence of differential routing. For the M0 microcontroller systems implemented in Chap. 4, no congestion issues were perceived.

To route the differential interconnect, controlled differential routing could be considered. Commercial EDA tools provide such routing, targeting differential clock nets or other sensitive differential signals. The tools are not optimized to route a full digital core with 20.000+ nets this way. The routing time increases signif-

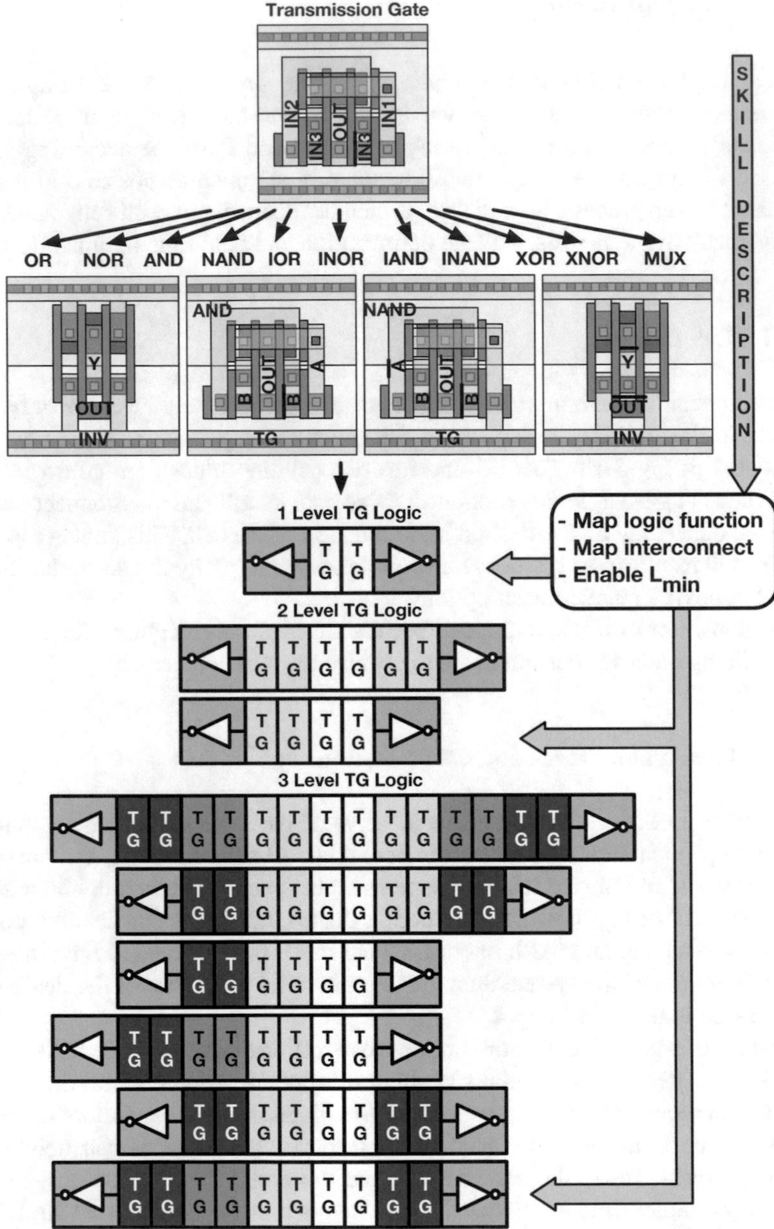

Fig. 3.18 A SKILL representation of the inverter and transmission gate is used to build a full standard cell library

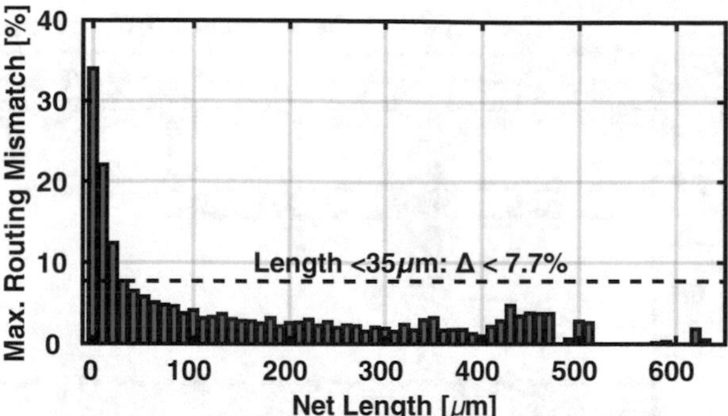

Fig. 3.19 Maximum routing length mismatch between differential pins in the microcontroller implementation

icantly under such restricted conditions. In the proposed design flow, differential nets are routed just as any other single-ended net. The tool does not consider the complementary counterpart of the net while routing. Figure 3.19 shows the routing results of the differential netlist in the M0 system. Only the shortest nets exhibit a larger relative routing length difference. The fact that differential nets always have their start and end pins in close proximity to each other aids to this. Because each differential pin has its own timing arc representation, the pin-specific physical effects (parasitic capacitance and resistance) are accurately taken into account.

3.4.3.3 Multi-Gate Length Leakage Optimization

One of the crucial steps in low-power physical implementation is leakage optimization. The proposed flow enables this by interchanging standard cells with different gate lengths. After routing, most of the physical implementation-dependent effects are present. The timing paths that exhibit leftover slack are slowed down by swapping fast leakage sensitive cells on the path with slower low-leakage cells. The process is identical to multi-V_T library optimization. In this work, the difference in V_T is omitted and different gate lengths are preferred. The motivation for this was discussed in Sect. 2.2.3. This leakage optimization is the main source for the improvement in energy consumption of the two microcontroller prototypes discussed in Chap. 4. As is shown there, multi-gate length leakage optimization results in a more than 4x leakage power reduction across the entire operating range.

3.4.3.4 Hold Time Optimization

A final step in the physical implementation process is hold time optimization. Hold time problems are an eyesore for low-power implementation. They increase area, active power and leakage power. Such timing violations are typically resolved by adding delay buffers on the short timing paths, avoiding faster propagation than the clock. While in normal conditions commercial EDA tools are perfectly capable of doing so, they exhibit some problems when handling differential logic. In normal conditions, the hold time optimization algorithm will investigate a timing path, place

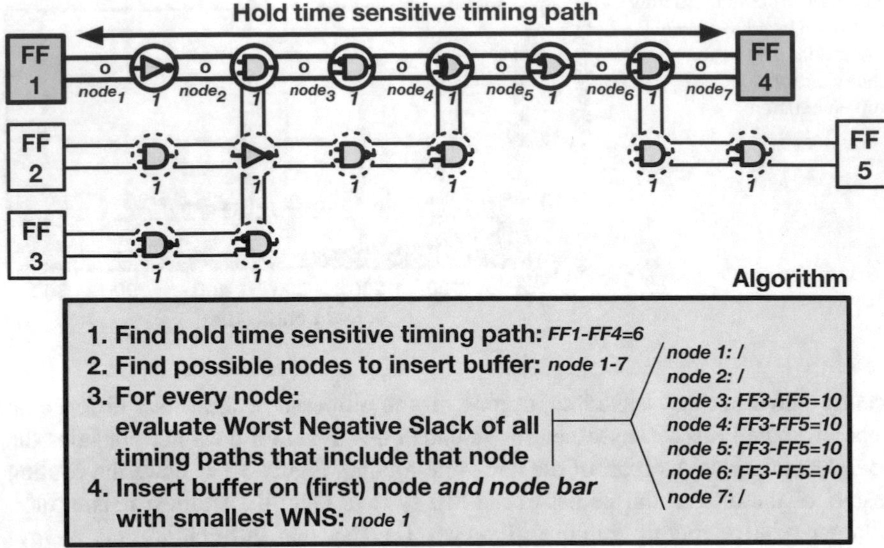

Fig. 3.20 Principle of the applied hold time optimization

the buffer and estimate the effect on hold time. Because the proposed flip-flops are differential, their hold time paths are not resolved by placing a buffer on a single net. The EDA tool hence concludes that placing a buffer is not beneficial for hold time optimization. Figure 3.20 demonstrates the technique used in this work. The actual hold time analysis and node selection is much like an EDA tool optimization. The crucial difference is that both the complementary and the regular path are padded in a single optimization cycle.

3.4.3.5 Limitations

While the proposed differential implementation flow is perfectly capable of delivering performant (microcontroller) systems, it has some limitations. As discussed, differential routing may introduce **congestion**, which could result in area overhead. The **hold time optimization** as presented here is effective, but does not employ the full power of the commercial EDA tools. A more important limitation is the **single-ended synthesis** and its effect on the physical implementation flow. In the proposed flow, synthesis is single ended and thus fully capable. Physical implementation is differential and is done by commercial EDA tools. These implementation tools comprise the same capability as a synthesis tool. Among others, they employ buffer adding, gate resizing, netlist restructuring, logic remapping, pin swapping, buffer deleting and moving instances to optimize the timing performance of the physical implementation. As the tools are not differential-capable, all synthesis-like operations are disabled. **Restructuring the netlist and remapping logic** is not possible for the differential physical implementation. Finally, the proposed flow did not focus on **testability** strategies such as scan chains and their insertion. If scan flip-flops are added to the standard cell library, these testability techniques can be added without much design overhead.

3.5 Proof-of-Concept

Throughout this chapter and the discussion of the different design flow steps, references to the 32-bit microcontroller systems implemented in Chap. 4 were made. Table 3.2 summarizes the application of the design flow details for these proof-of-concept implementations. Specifics of the used standard cell libraries are shown,

Table 3.2 Application summary of the proposed design flow on a proof-of-concept microcontroller system of Chap. 4

	ULV differential	$V_{dd,IO}$ library
Library	300 mV, SS-TT-FF	900 mV, SS-TT-FF
	467 cells, L = 40–60 nm	Standard CMOS top. & 6T SRAM
	Differential TG logic	IO logic, macros, IO pads
	Single-ended clock buffers	Normal clock tree
	Level shifters/domain shifters	Level shifters
Extra	ULV *Pseudo* Single Ended	/
	Deduced from ULV differential	

	32-bit ARM Cortex-M0 system		
Design	Core	Peripherals	Memory
	Power domain 1	Power domain 2	Power domain 3
	ULV domain		IO domain
	Differential		Single Ended
	151 level shifters		

	Action	Effect
Implem.	Single-ended synthesis + verification	$\Delta = 15\%$
	Differential gate-level netlist + diff. verification	
	Self-inverting technique	−35% logic gates
	Clock mesh	/
	Differential routing	Minimal Δ route
	Multi-gate length	$\div 4$ P_{leak}
	No netlist restructuring or logic remapping	/

	Spec.	Result
Silicon	Voltage	370 mV
	Speed	13.7 MHz
	$E/cycle_{core}$	8.80 pJ
	$E/cycle_{periph.}$	2.77 pJ
	$E/cycle_{memory}$	31.65 pJ

as well as design and implementation details and silicon measurement results. When considering the standard cell libraries, three libraries are used: the ULV differential, the ULV *pseudo* single ended and the $V_{dd,IO}$ (part 1). This maps nicely to the system functionality, where the core and peripheral logic are implemented in the ULV differential domain, while the memory uses the $V_{dd,IO}$ domain (part 2). Physical implementation requires different steps, each of which influences system performance (part 3). Finally, the silicon implementation results in state-of-the-art measurement results (part 4).

The design flow as presented here clearly delivers on its functionality. The applied scripts and tools are generic and can be applied in other ways. When the library and gate-level netlist are loaded in a database, it becomes easy to implement scripts that adapt and augment the netlist and the tool flow. An example of such an augmentation is the error detection and correction system presented in Chap. 6. There, timing analysis is used to select the most critical paths of the design. The endpoint flip-flops of these paths are replaced by error detection and correction flip-flops using similar scripts as used in this chapter. Additionally, the *error* signals these flip-flops generate are propagated through the gate netlist hierarchy to the system-level error processor, which incorporates these in its control loop. Scripts that enable such operations are easily implemented in the proposed design flow. More details on this augmentation are presented in Sect. 6.3.

3.6 Conclusion

This chapter sketched the VLSI design strategy for differential transmission gate logic. Typical VLSI design flows rely on well-considered partitioning of the design cycle. RTL descriptions, logic synthesis and standard cell implementation are the industry standard. The proposed design flow augments the typical design strategy by selectively intervening at different levels in the hierarchy. At the **library level**, this chapter introduced differential transmission gate-based standard cells. The different representations of these standard cells are generated, analysed and summarized in a fully differential standard cell library. At the **logic level**, the proposed flow enables single-ended synthesis using differential logic. By compiling a *pseudo* single-ended standard cell library from the original differential standard cell library, the synthesis tool can use single-ended logic to synthesize the RTL description. The generated single-ended gate-level netlist is transformed to the differential domain, in accordance with the original differential standard cell library. At the **physical level**, the low-power implementation flow is used to implement a performant system. A dedicated power domain is used for the ULV differential logic. As such, the differential system can interact with single-ended (higher-voltage) logic or single-ended macros.

In this process, the design flow delivers on its desired functionality. It enables **standard cell design** similar to the industry standard. By using commercial tools, the design flow is relatively **easy and fast**. Low-power implementation techniques such as multi-mode multi-corner optimization are necessary to keep **design margins** at their lowest while enabling **ultra-low-voltage operation**. With standard cell libraries comes timing and power analysis. Accurate **speed predictions and power footprints** can be extracted with much less effort than in a full custom design flow. Doing this analysis for multiple supply voltages provides insight in the static and dynamic power consumption, indicating the target region for **minimum energy operation**.

The VLSI design flow is put to practice in Chap. 4. It implements two prototypes of a 32-bit microcontroller system. The outlook in Sect. 3.5 gave an idea about the applicability of the different design steps. Chapter 4 will demonstrate that the proposed strategy succeeds in delivering a **relatively large digital system** with state-of-the-art speed and energy performance. The conclusions presented in this chapter were published in [5, 6]

References

1. Cadence: SKILL Language User Guide. Cadence User Guide **03** (2015)
2. Gajski, D., Kuhn, R.: New VLSI tools. IEEE Comput. Soc. **16**(12), 11–14 (1983)
3. McGuire, P.: Pyparsing—An Alternative Approach to Creating and Executing Simple Grammars. http://pyparsing.wikispaces.com/
4. Power Forward: A Practical Guide to Low-Power: Design User Experience with CPF, pp. 62–63 (2009)
5. Reyserhove, H., Dehaene, W.: A differential transmission gate design flow for minimum energy sub-10-pJ/cycle ARM Cortex-M0 MCUs. IEEE J. Solid State Circuits **52**(7), 1904–1914 (2017)
6. Reyserhove, H., Dehaene, W.: Design margin elimination in a near-threshold timing error masking-aware 32-bit ARM Cortex M0 in 40nm CMOS. In: 43rd IEEE European Solid-State Circuits Conference (ESSCIRC), pp. 155–158. IEEE, Leuven (2017)
7. Sasao, T.: Logic Synthesis and Optimization. The Kluwer International Series in Engineering and Computer Science. Springer, New York (1993)
8. Synopsys: Liberty User Guides and Reference Manual Suite v. 2014.09. Tech. rep.
9. Weste, N., Harris, D.: CMOS VLSI Design: A Circuits and Systems Perspective, 4th edn. Addison-Wesley Publishing, Boston (2010)

Chapter 4
Ultra-Low Voltage Microcontrollers

Abstract To leverage the design strategy presented in Chap. 3 to its full capability, this chapter implements a state-of-the-art microcontroller system. The main goal is to provide a proof-of-concept implementation on an industry-proven design to showcase the efficacy of the mentioned strategy, as well as achieve excellent energy and speed performance. The ARM Cortex-M0 core is chosen to this end. It is used in a variety of commercial systems going from IoT nodes (Sparkfun, NEST: Nest Thermostat Teardown. https://learn.sparkfun.com/tutorials/nest-thermostat-teardown- (2016); TechInsights, Fitbit: Fitbit Charge 2 Teardown. http://www.techinsights.com/about-techinsights/overview/blog/fitbit-charge-2-teardown/) to virtual reality (VR) glasses (IFixit, Oculus-VR: Oculus Rift CV1 Teardown - iFixit. https://www.ifixit.com/Teardown/Oculus+Rift+CV1+Teardown/60612, 2016). The core is ideally suited for energy-constrained applications, so it has been a recurrent topic in low voltage literature as well.

This chapter will demonstrate how the architecture of the microcontroller system came to be in Sect. 4.2. It includes the M0 core, a memory and the necessary peripherals for interfacing and debugging. The framework to program the system is fully compatible with the ARM tool chain. It allows arbitrary C-code to be programmed and run on the core. Section 4.3 dives deeper in the implementation details of the presented system. Two distinct silicon implementations of the same system have been realized. A number of considerations necessary for ultra-low voltage operation are discussed. The measurements presented in Sect. 4.4 demonstrate the state-of-the-art performance of the prototypes. They excel in speed performance and ultra-low energy consumption. The memory system power and the overall standby power leave room for improvement. Section 4.5 provides more details on the energy consumption and how it is improved between both prototypes. The state-of-the-art comparison in Sect. 4.6 further demonstrates the good performance of the prototypes. Finally, some concluding remarks regarding the system and its performance can be found in Sect. 4.7.

H. Reyserhove, W. Dehaene, *Efficient Design of Variation-Resilient Ultra-Low Energy Digital Processors*, https://doi.org/10.1007/978-3-030-12485-4_4

4.1 Introduction

A lot of applications come to mind when thinking about low voltage low energy computations.

- A wireless sensor node gathers sensor data and often does some pre-processing to compress the information before data transmission [41], often referred to as *edge computing* [30].
- Monitoring of (bio-)medical parameters on the human body for, e.g., ECG usually relies on algorithms with a decent amount of processing to extract the medically relevant information [5, 19].
- A nanodrone runs a control loop using gyroscope and accelerometer data to stabilize itself and can use computer vision image processing to navigate [32].
- An NFC or RFID enabled application requires some cryptographic calculations to create go/no-go information [8].
- Hearing aids rely on complex DSP processing to localize and filter spatial audio information [13].

All these applications put heavy constraints on three conflicting aspects of system design: the available energy, the available computing capability and the device form factor (see Fig. 4.1). The microcontroller, being at the heart of all of these systems, influences these aspects heavily by running the sensor processing, control loops or overall system management. The ARM Cortex-M0 provides a highly performant 32-bit instruction set with minimal power, area and code footprint, ideally suited for these operations. A low energy high-speed implementation of this microcontroller can improve these systems significantly and enable new applications with even tighter constraints.

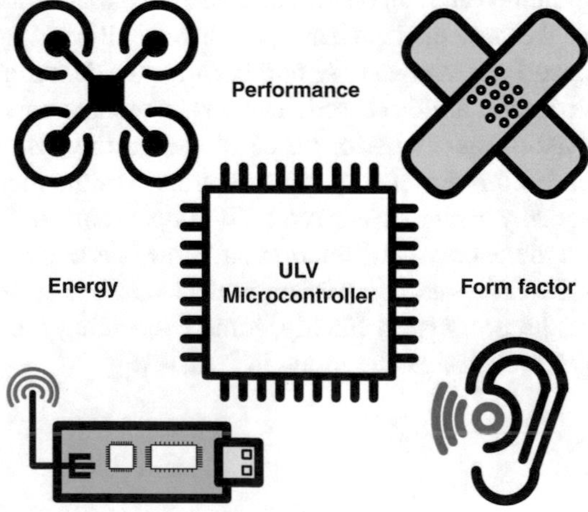

Fig. 4.1 Applications in the ultra-low voltage microcontroller domain constrained by three key factors: energy, performance and form factor

4.2 Architecture

A typical microcontroller system is shown in Fig. 4.2. It consists of the core processor, memory, sensor interfacing and a communication link. The core can be a Cortex-M0 or similar. The memory usually consists of a volatile part (SRAM) and a non-volatile part (ROM and flash). Connectivity can be done through I2C, SPI, UART, USB, or a direct wireless link like Wi-Fi or Bluetooth. Analog, sensing and control blocks can help interfacing through A/D and D/A converters, comparators and PWM generators. The exact implementation of commercial systems can vary between vendors or flavour. In this work we limited the system to the core processor, the memory and a limited number of interfacing blocks to enable simple benchmarking and debugging. The details of this implementation can be found in the next subsections.

4.2.1 Microcontroller System

The microcontroller system presented in this section is the base for the silicon implementations further in this chapter and in Chap. 6. A block diagram is shown in Fig. 4.3. Its main focus is the ARM Cortex-M0 core. The necessary blocks were added to facilitate operation, debugging and simple interfacing with the goal of enabling reliable silicon measurements and potential problem assessment. An SRAM memory functions as start-up ROM, instruction and data memory. A bus implementation allows memory-mapped communication between all the blocks. The core acts as the primary bus master, but can be bypassed by the debugger to

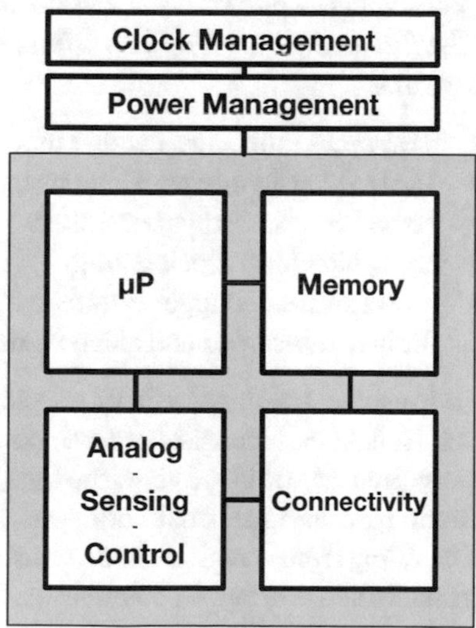

Fig. 4.2 Typical microcontroller system and subdivision in its most important building blocks

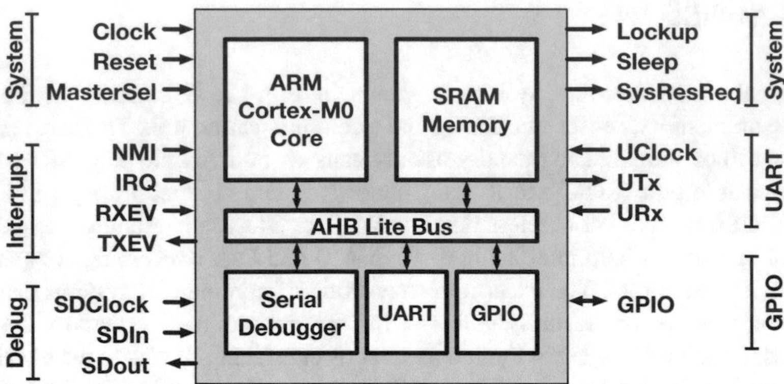

Fig. 4.3 Overview of the full microcontroller system with interfacing

extract information at any time during execution. Apart from the hardware interrupts available on the core, a UART interface and 8 GPIO ports allow interfacing with the system.

4.2.1.1 ARM Cortex-M0 Core

The ARM Cortex-M0 initially differentiated itself in the market because it enabled 32-bit instructions and data in a ultra-low-power small area core. This provided the performance efficiency for new applications and faster design time, also because of its good C-code compatibility. A comprehensive overview of the core can be found in [43]. This work equips a *lite* version of the core, intended to realize fast silicon prototyping as part of the ARM Cortex-M0 DesignStart University program [4, 12]. The *lite* version (see Fig. 4.4) differs from the full version in the following (optional) features:

- a 32-cycle multiplier instead of a single cycle multiplier.
- 16 fixed hardware interrupts instead of the programmable 1–32.
- no wake-up interrupt controller.
- no architectural clock gating.
- no hardware debugger interface and no hardware debug support.
- limited low-power and sleep signalling.

Despite this, the *lite* core is fully compatible with the ARMv6-M Thumb instruction set. Its data-path consists of three pipeline stages: fetch, decode and execute. From a programmer's point of view, the core switches between two operation modes: thread mode (executing normal code) and handler mode (executing exception handling). Operating from a register bank consisting of a number of general-purpose registers, pointers and a program counter, the M0 realizes in-order execution of the machine code program in the memory.

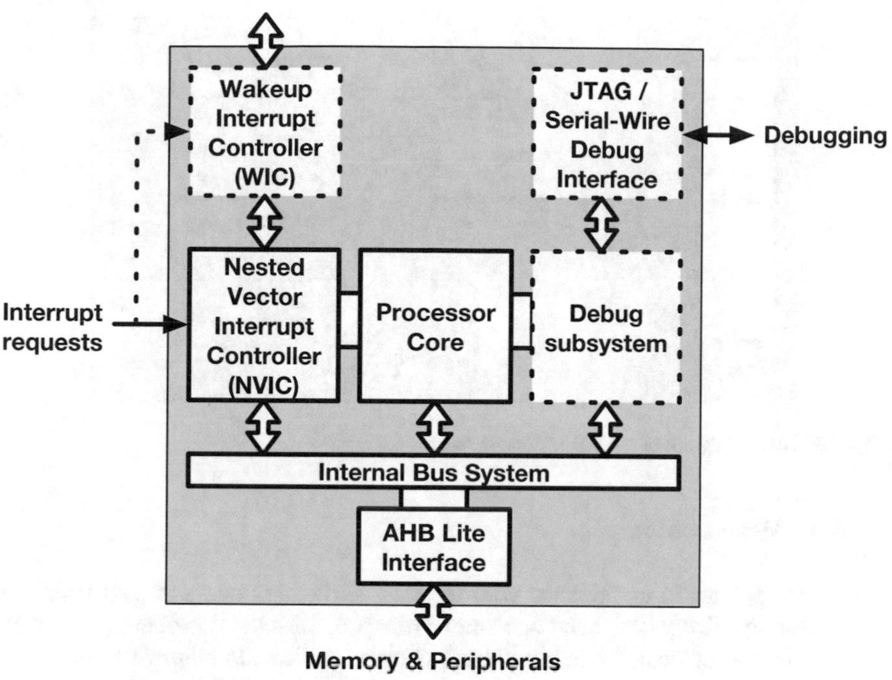

Fig. 4.4 Block diagram of the M0 core internal subsystems. Optional subsystems not present in the core in this work are dash-lined

4.2.1.2 AHB Lite Bus

The core interfaces with other blocks using the advanced microcontroller bus architecture (AMBA3) AHB-lite (advanced high-performance bus lite) protocol. It is a master–slave bus protocol where the M0 core acts as a master, while all other blocks (memory, UART, GPIO, etc.) act as slaves. The main idea is very simple: Fig. 4.5 shows a single master controlling the address and write data of all slaves while selecting read data from a single slave according to the applied address. Other than the address and data bus, additional control signals facilitate burst transfer, locked transfers, protection control, different data sizes and transfer completion or extension. Every bus transaction consists of a one cycle address phase and a one or more cycle data phase. An example can be seen in Fig. 4.12. The full protocol details can be found in [2]. Important to know is that every bus master or slave requires a bus controller to handle the different cases of data transfer. Moreover, the AHB bus targets high performance or high data rate blocks of the system. Lower performance peripherals are usually served by a slower APB (advanced peripheral bus) bus system, acting as a separate subsystem of the AHB bus.

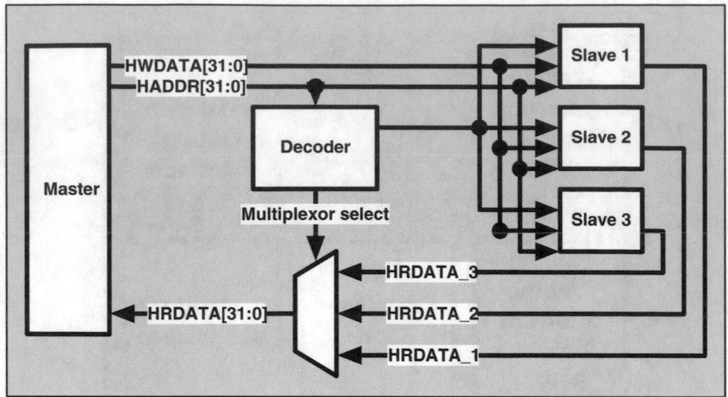

Fig. 4.5 Block diagram of a AHB-lite bus system

4.2.1.3 Memory Map

The core operates in compliance with the bus, retrieving code and data from the main memory. Every peripheral is memory-mapped, meaning it is operated through the bus with a dedicated address. The M0 uses 32-bit addresses with byte level addressing, which results in an address space of 4 GB of addressable data. The M0 architecture prescribes a number of regions with a recommended use to help software porting between devices. The prescribed memory stack is shown in Fig. 4.6 together with the memory map equipped in this work. While large address spaces are used for devices like the UART or GPIO, only a few registers are used. Employing the full address range for these few registers allows easy bus arbitration and helps prevent functional faults in the processor code.

4.2.1.4 Memory

As can be seen in the memory map, only a small range of the address space is mapped to actual memory. Commercial microcontroller systems equip a hybrid memory system consisting of non-volatile and volatile memory. The boot loader program resides in a factory programmed ROM, which cooperates with on-chip SRAM memory and on- or off-chip flash memory. Program code is often stored in flash due to its non-volatility, while the faster SRAM is used as the primary memory for the stack and data storage. The design of the memory system of a commercial microcontroller includes considerations such as memory size, speed, power-up time, reliability, security and many others.

In this work, the memory system is limited to a single on-chip SRAM. The SRAM is used to boot the system and contains the stack as well as the program code and data. A *linker* file (see Fig. 4.7) is used during compilation to control the memory layout of the output file and map it correctly to the final image. Since the

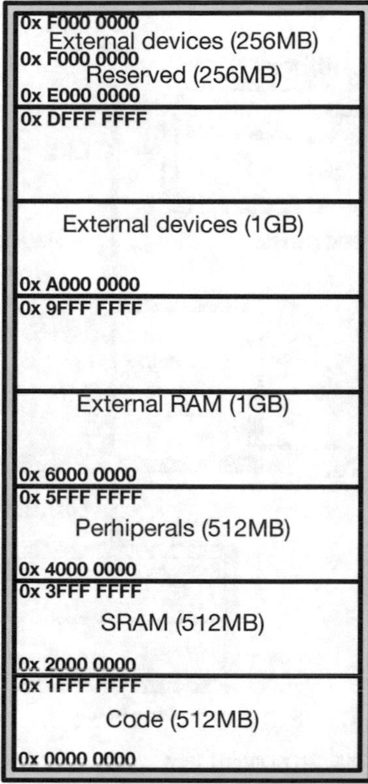

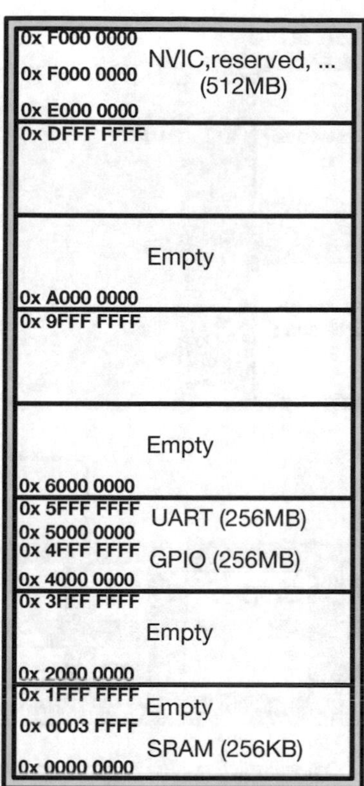

Fig. 4.6 Memory map prescription of the M0 architecture (left) and equipped memory map in this work (right)

SRAM is volatile, the memory content has to be re-written to the SRAM after each power-up. While this would be impractical for a system in the field, it is convenient for silicon testing. The boot program, memory layout and appointed memory space can be adapted at will if the test program requires it. To guarantee enough available memory for operation, 256 KB (prototype 1) or 64 KB (prototype 2) on-chip SRAM memory was equipped. Commercial systems use similar size on-chip memories, but use flash memory for the bigger chunk. Additional memory can be added off-chip through a (slower) serial link (UART, SPI and I2C).

4.2.1.5 Debugger

The ARM Cortex-M0 architecture provides a debugger subsystem in the core. Functionality such as program breakpoints and data watchpoints is possible through this system. When such a debug event occurs, it can put the processor core in a halted state so the programmer can evaluate the processor state at that exact point

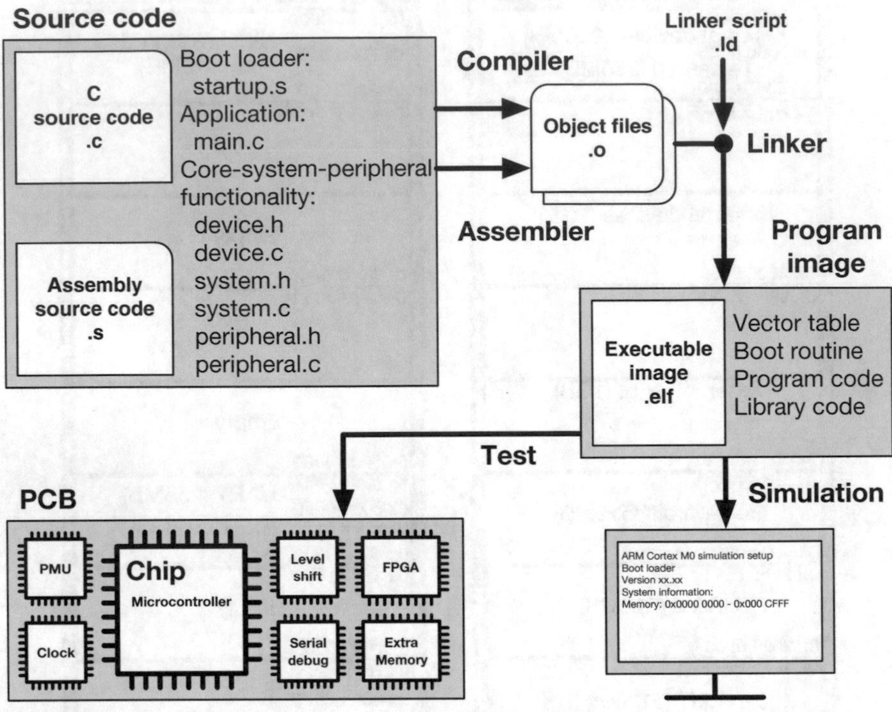

Fig. 4.7 Typical microcontroller program development and mapping flow

in execution. Additionally, a debug interface allows access to the bus system and the debugging functionalities. In the past, the JTAG-protocol was very popular for interfacing with this debugger. Recently, the two wired serial wire protocol is frequently being used.

In this work, the ARM core is not equipped with the debug functionalities as just described. However, a debugger is implemented to communicate on the bus, interface with the bus slaves and read or write the memory. In the presented setup the debugger is crucial to the microcontroller functionality. It programs the memory with the program code and accompanying data. During test it can also be used to control the other peripheral systems, as well as read out specific memory addresses at any time while the program is being executed.

4.2.1.6 Interrupts

Interrupts or interrupt service routines (ISR) are crucial to the operation of a microcontroller system. In an application, they represent real-time events occurring through interaction with other (off-chip) systems such as sensors, buttons and communication links. The ARM Cortex-M0 core inherently provides up to 32 hardware interrupts. They are managed by the nested vectored interrupt controller (NVIC): a controller which gathers and senses interrupts, considers their priority and masking and runs the accompanying C-code which is to be executed in case of the interrupt. The non-maskable interrupt signal (NMI) is a special hardware interrupt which cannot be ignored by normal interrupt masking techniques. Commercial systems often add a wake-up interrupt controller (WIC) to recover the system from sleep or deep sleep using interrupts.

The interrupts in this system are used both internally and externally. Internal interrupts are used by the UART and GPIO to signal events to the core, e.g., the arrival of a new byte on the UART bus or GPIO ports. External interrupts are usefully employed in the test setup to trigger certain parts of test code in varying conditions.

4.2.1.7 UART

The universal asynchronous receiver–transmitter (UART) is historically probably the most common peripheral in microcontroller systems (apart from the memory). It enables asynchronous serial communication with configurable data format and transmission speed. Almost all computing, communication and I/O devices provide a serial communication link compatible with UART hardware. The UART is a relatively simple block that could not escape the microcontroller system of this work.

4.2.1.8 GPIO

Apart from the interrupts and the UART, the general-purpose input output ports (GPIO) are the third option to interface with the microcontroller of this work. These ports are generic pins available on bond pads of the silicon implementation or pins of the test board without a predefined purpose. Their direction (input or output) and behaviour are controllable by the user program at runtime. The idea is that the system designer can use these pins as digital control lines in his system. Similar to the other interfaces, the 8 GPIO ports equipped in the microcontroller presented in this work are useful for system interaction, testing and debugging.

4.2.2 Microcontroller Framework

In most applications the microcontroller runs a C-code program while interacting with peripheral systems. While some parts of the program may be system specific, most of it usually is generic C-code. The bigger part of the program is mapped to the hardware implementation in a development flow, of which the key components are the compiler/assembler and the linker. An overview of the full flow can be seen in Fig. 4.7. The final program image consists of the vector table, the boot routine, the program code and the necessary library code. This program image is used to run a simulation or a silicon implementation after loading it in the microcontroller memory using a test board.

4.2.3 Simulation and Testing

A good simulation strategy is necessary to test the functionality and performance of a relatively complex system such as a microcontroller. Throughout the entire design flow, implementation and measurement, a single test-bench is used. It covers the basic functionality of the debugger, each of the peripherals and the core. Simulations are performed identically to the targeted measurement: the debugger programs the microcontroller memory from a program image, after which the microcontroller is taken out of reset and initialized. Different parts of the test code are triggered using interrupts. Feedback regarding the correct operation can be seen through direct interaction with the peripherals, as well as by reading out specific addresses with the debugger. This can be done at run-time by freezing the core clock separately from the bus and peripherals.

4.2.4 Benchmarking

While a test program covering the functionality of the system is necessary to verify whether the system actually works, a lot of the test metrics depend on the application or the program. A program representative of in-field operation can be used to analyse system functionality, speed performance and energy consumption. While more recent benchmarks like CoreMark or SPEC have grown popular, the Dhrystone benchmark C-code program [42] is still a good and easy match for this work. The program does not compute anything meaningful, but it is syntactically and semantically correct. It balances assignments (51%), control statements (32.4%) and procedure/function calls (16.7%). First and foremost, the Dhrystone benchmark is a performance benchmark. A typical result for a Cortex-M0 is 0.89 DMIPS, with DMIPS defined as the number of Dhrystones (loop iterations of the program code) executed per second, divided by 1757. Since the compiler maps the Dhrystone code to a fixed number of instructions, the benchmark depends mostly on the (allowed) compiler options and the operating speed of the core. Typically, compiler options such as function inlining are not allowed when compiling the benchmark (or such options should be reported carefully). More interestingly, in this work we report energy metrics while running the Dhrystone program. The benefit of this approach is that circuit activity (which is highly related to the energy consumption) is fixed and comparable with other implementations running the same program.

4.3 Implementation

This section presents the different implementation details of the microcontroller system shown in Fig. 4.3. Two distinct silicon implementations of the same architecture are presented in this and the following sections: *prototype 1* and *prototype 2*. Their biggest difference is the minimum energy optimization in going from the first to the second prototype. This is done through a different device sizing in the logic cells, a supplementary low-leakage logic library and a smaller SRAM memory. The key differences are highlighted again in Table 4.1. Further in this section, we focus on the RTL of the system, the timing sensitive aspects and clock, the power distribution, synthesis and simulation results.

4.3.1 RTL

The bulk of the full system RTL code is taken up by the ARM Cortex-M0 core. It was obtained through the ARM DesignStart University Program [4, 12]. At the time of design, this meant the verilog code was obfuscated. The code can be simulated and synthesized, but it is nearly impossible to distinguish what register or signal serves which purpose. In that sense, it is very similar to flattened netlist, as in without hierarchy, intuitive name calling or structure. While this forced us to approach the design generically, and thus speaks to the applicability of the proposed design flow and library, this also results in a number of limitations. There is no knowledge of the pipelined subdivisions and architectural optimization or intervention is difficult. It is also quite difficult to segment the design in multiple power domains intended for semi-power down, sleep or deep sleep modes. Efficient clock gating is also difficult without detailed knowledge of the core. For this reason, most of these techniques are not used in this chapter. While

Table 4.1 The two microcontroller implementations of this work with their key differences

	Prototype 1	Prototype 2
System	ARM Cortex-M0	ARM Cortex-M0
Peripherals	Debugger, UART, GPIO	Debugger, UART, GPIO
Memory	2× 128 KB SRAM block	1× 64 KB SRAM block
	0.9 V $V_{\mathrm{dd,nominal}}$	0.9 V $V_{\mathrm{dd,nominal}}$
Library	ULV differential	ULV differential
	40 nm gate length	40 nm/60 nm gate length
	Optimized for $V_{\mathrm{dd}} = 250\,\mathrm{mV}$	Optimized for $V_{\mathrm{dd}} = 300\,\mathrm{mV}$
Self-invert technique	No	Yes
Slow–slow sign-off	0.4 MHz, 250 mV	0.7 MHz, 300 mV
Power domains	3	2
	Core, peripherals, memory	Core logic, memory

this definitely impacts the achieved results and figure-of-merit, achieving good results without these techniques highlights the proposed approach. There is no doubt optimizations can be done. However, there is also no doubt a RTL team from a big microcontroller vendor would do it even better. Myers et al. [29] show how a fine-grained power domain division (it uses 14 power domains) can greatly benefit the active and standby energy consumption of an ARM Cortex-M0+ system. Apart from the core, the AHB bus arbiter and peripherals are custom coded according to the desired functionality and bus protocol. Cobham Gaisler [10] provides an open-source IP library that forms a good starting point for such peripherals.

4.3.2 Voltage and Power

The microcontroller system is divided in two voltage domains: an ultra-low voltage domain for the logic and an I/O voltage domain for the memory and chip interfacing and signalling. The voltage domains result in three power domains for prototype 1 and two for prototype 2, since prototype 1 has a separate power domain for the core peripherals. The standard cell design flow uses a power intent file (CPF) to formally structure the different power domains, voltages, libraries and low-power cells. Figure 4.8 shows the power domain subdivision and location on the chip floor plan for prototype 1. The ULV domain targets a 250 mV V_{dd}. The memory and I/O power domain operate at the nominal supply voltage of the technology (900 mV). Due to this voltage difference, level shifters are inserted between both domains. The power intent file makes the design flow aware of this by specifying separate level shifters for every domain crossing. In this work, the ultra-low-power domain also operates with differential signals. By coinciding the ultra-low-power domain with the differential domain, the level shifters provide double functionality: voltage level shift and differential to single-ended domain. This technique was well discussed in Sect. 3.4.2. The power grid is layed out across the entire floor plan to provide a dense power delivery and minimal IR drop. An early power rail analysis on the floor plan showed a maximum IR drop of less than 10% across the design.

4.3.3 Timing

The microcontroller system is a fully synchronous system controlled by a single clock signal. The low voltage operation clock distribution requires careful consideration to result in an energy-efficient system. The most important metrics to this end are the clock transition time, skew and insertion delay:

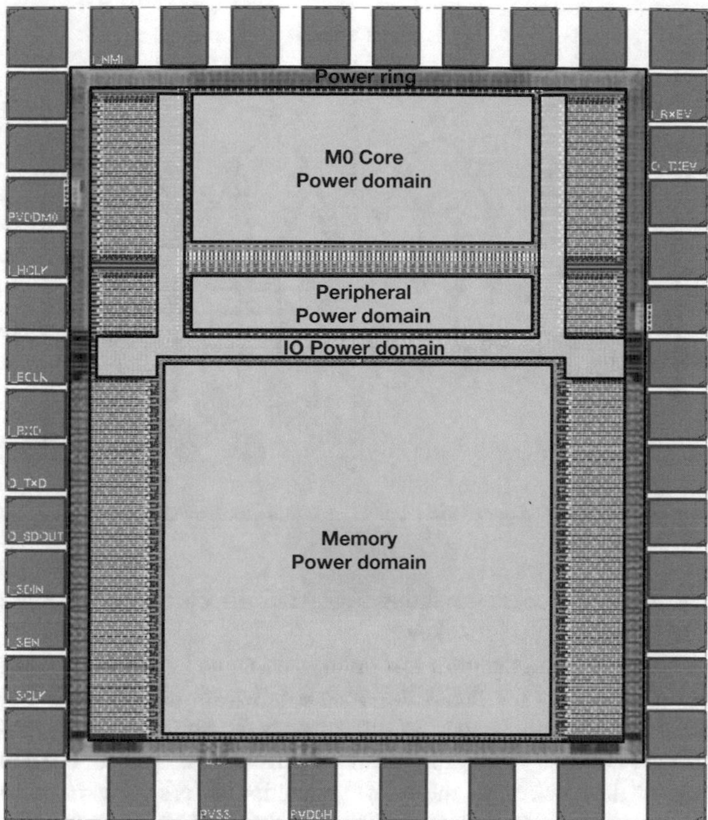

Fig. 4.8 System floor plan with power domain subdivision and power grid

- **Transition time** is defined as the time the local clock signal needs to transition from high to low or vice versa. Slow transition times can occur due to a clock buffer not being able to charge its output capacitance fast enough.
- **Skew** is the difference in clock arrival time at different endpoints in the clock tree. In a static timing analysis, identical buffering of separate clock tree branches will lead to minimal skew. Mismatch between the clock buffers results in different arrival times and thus skew. Note that ultra-low voltage cells are highly susceptible to mismatch.
- **Insertion delay** or clock source latency is defined as the time it takes for the clock to propagate from its source to the clock definition point, e.g., a power domain crossing.

Seok et al. [36] discuss the details of low voltage clock distribution at length. A key observation is that low voltage clock distribution does not benefit from extensive and repetitive clock buffering to overcome RC delay. The clock buffers are typically quite slow compared to routing due to their low drive current. Moreover, they are

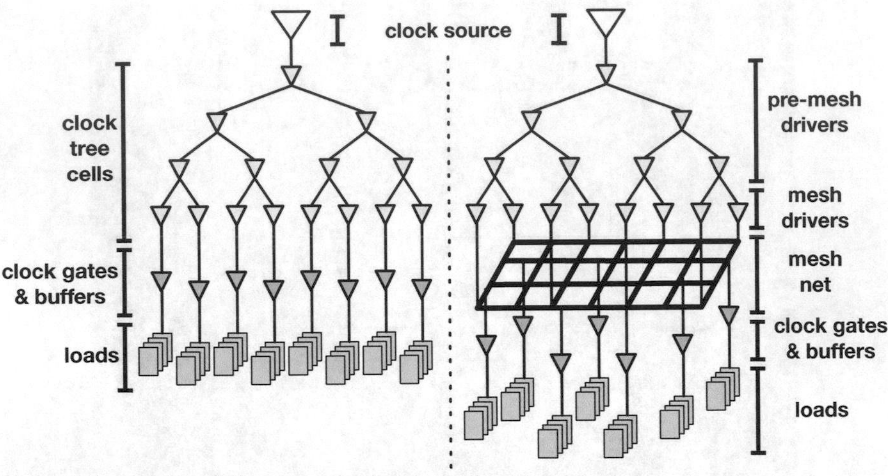

Fig. 4.9 Comparison between conventional clock tree and clock mesh

highly susceptible to process variations. These two effects typically result in a high insertion delay and a high clock skew.

To battle these effects, the ultra-low voltage domains in this work equip a clock mesh. Figure 4.9 shows the major differences between a conventional clock tree and a clock mesh. The most important difference is the mesh net. It is a multi-driven (redundant) net attached to an array of mesh drivers. The goal is to smoothen the arrival times of the clock at the mesh level, resulting in less skew. In essence, **only the unique parts of the launch and capture paths add to the skew**. Because the mesh is redundant, it contains much less unique parts than the conventional clock tree.

To further battle mismatch and insertion delay, and drive the large capacitance of the mesh net, sufficiently large buffers are used as pre-mesh and mesh drivers. An overview of the clock mesh of the M0 core is depicted in Fig. 4.10. A similar structure is used for the peripheral power domain of prototype 1. During clock tree synthesis, the place-and-route tool extracts a SPICE netlist of the clock mesh to get more accurate simulation results. Figure 4.11 shows the clock signal across the full hierarchy of the microcontroller system. Each power domain has its own clock signal. Across power domains, a dedicated level shifter is used for the clock signal. Because the clock level shifters and clock mesh drivers operate at ultra-low voltage, they have significantly more delay than the clock buffers in the I/O domain. A large difference between both clock signal insertion delays can be perceived. Clock tree synthesis aims to match clock arrival times across the system. In a mixed voltage domain system such as this, the tool naively tries to mimic the ultra-low voltage clock delay by adding a lot of clock buffers in the I/O domain. This leads to huge overhead in the I/O domain. In this work, we overcome the insertion delay mismatch by re-inserting the ultra-low voltage clock in the I/O domain. This way, the I/O domain clock shares the large clock insertion delay of the ultra-low voltage domain.

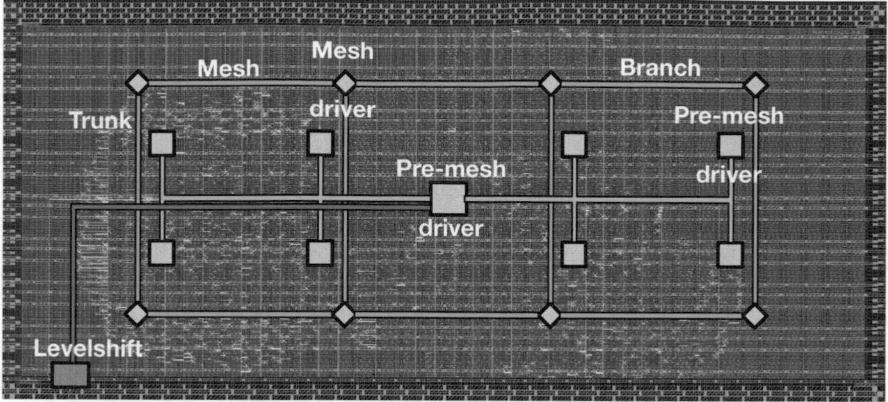

Fig. 4.10 Clock mesh implementation of the ARM Cortex-M0 core

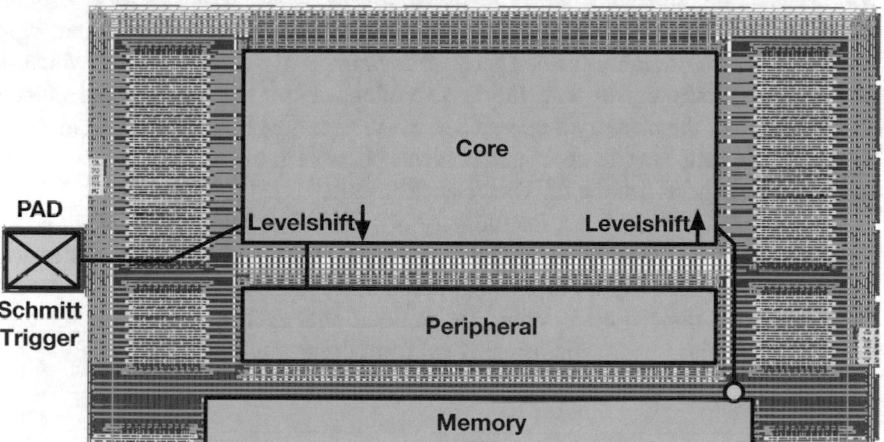

Fig. 4.11 Overview of the full system clock hierarchy

The same principle as before holds here: a larger common part in the clock path results in less mismatch.

Additionally, a major adjustment is the memory being clocked on the negative clock edge. While this puts tighter timing constraints on the memory, it significantly improves the timing relationship between the ultra-low voltage logic and the memory. Operating the memory on the negative edge while the logic operates on the positive clock edge breaks the hold time constraint between the two. In any other case, the hold time constraint would be hard to meet due to the large voltage difference and clock skew. The hold time optimization as discussed in Sect. 3.4.3 is used to pad the differential short paths in the design. Synthesis and place-and-route results are discussed in Sect. 4.3.5.

4.3.4 Memory

Memories form an important part of any microcontroller system. A significant part of the overall energy consumption can be attributed to them. Because the memory is typically a large circuit with a low activity, it requires a different consideration than ultra-low voltage logic. The speed performance of a nominal V_{dd} memory is typically one or more orders of magnitude higher than an ultra-low voltage logic block. Rooseleer et al. [35] and Sharma et al. [37] highlight some of the key aspects in designing a memory for applications such as microcontroller systems. A key point is to prioritize energy efficiency over speed and silicon area. While a state-of-the-art microcontroller system definitely requires an energy-efficient SRAM memory, this work operates a standard SRAM block provided by the technology foundry. It does not prioritize energy efficiency and thus is overdesigned for the targeted application. The choice for such an SRAM is motivated by practicality: a wide range of SRAMs was available on demand, while a low-power SRAM would require a lot of custom work. The foundry SRAM also resulted in the power domain split between logic and memory: the memory block operation is warranted down to 0.81 V which is two to three times higher than the logic voltage. Note that in the final silicon implementation, the memory is operated at 0.6 V. Although this is not recommended by the SRAM datasheet, correct operation was possible at this voltage across a wide temperature range and at the full speed specification.

As discussed in Sect. 4.3.3, the memory is operated at the negative clock edge. Figure 4.12 shows the detailed operating principle of the AHB bus in combination with the SRAM memory for read and write operations. Due to the AHB protocol write operation, the address is stored for an additional cycle before it is applied to the SRAM address bus. A single cycle read from the memory is possible.

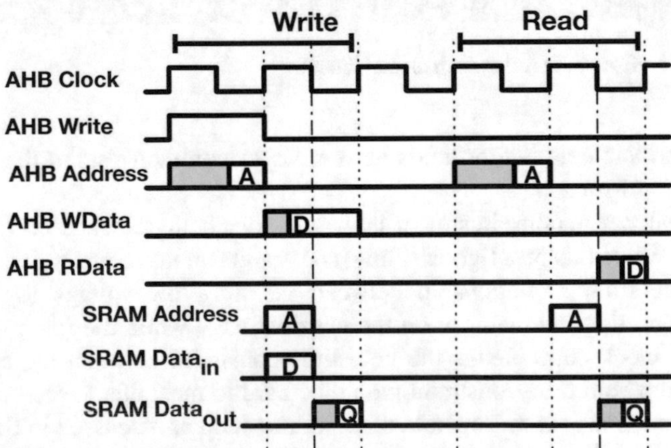

Fig. 4.12 Bus and memory read and write operation on the negative edge of the clock in relation to the AHB protocol

4.3.5 Synthesis and Place-and-Route

The synthesis flow as described in Chap. 3 is used to implement the full microcontroller system. The RTL code is synthesized using a single-ended library extracted from the differential library. After functional verification, the tool flow remaps the gate level netlist to the differential library and timing is verified. By relaxing the timing constraint on the design, logic gates and area can be traded for speed performance (see Figs. 2.28 and 2.29). Additionally, the synthesis script is instructed to put a lot of effort in leakage power optimization. For this, it uses path activation data from the Dhrystone C-code RTL simulation.

Table 4.2 shows the logic mapping results after synthesis for both prototypes. Both prototypes consist of a little over 1300 flip-flops and around 10,000 logic gates. Prototype 1 has a significant amount of inverters. Prototype 2 benefits from the self-inverting technique presented in Sect. 3.4.2.4 and removes all inverters. Both prototypes benefit from the complex logic gates available when using 2- or 3-level transmission gates: in both cores a large part of the functionality is implemented in 2- or 3-level TG (42% and 80%). To give an idea of the complex functionality gates, the most used transmission gate logic gates in the core are (in order): OR22NAND3, AND22NOR2 and OR3NAND2 (3-level TG), OR22NAND2, AND2NOR2 and OR2NAND2 (2-level TG) and NOR2, OR2 and NAND2 (1-level TG). More details were presented in Fig. 3.15. Prototype 2 improves on prototype 1 even further by employing a large number of long-gate length cells (close to 75% in the core), which trade off better leakage power for a slightly reduced speed performance.

Table 4.2 Synthesis logic mapping results of the ARM Cortex-M0 system

	Prototype 1		Prototype 2	
	Core	Peripherals	Core	Peripherals
Flip-flops	841	491	841	534
Logic gates	9842	1367	6383	1641
Inverters	3486	408	0	0
TG 1-level	3660	652	1295	525
TG 2-level	2360	218	4066	720
TG 3-level	336	89	1022	396
Low-V_T 40 nm cells	100%	100%	25.9%	68.6%
Low-V_T 60 nm cells	0%	0 %	74.1%	31.4%
Levelshifters	90		61	

4.3.6 Simulations

To get accurate simulation results of the speed, power and leakage percentage of the ULV systems, a sign-off accuracy full-chip simulation of the system is done in a dedicated power analysis tool. The simulation incorporates the activity of the circuit by using a real test program. Moreover, the power analysis is preceded by an accurate parasitic extraction. Without accurate simulations and parasitic extraction the power analysis results differ so significantly from the measurements to the extent that they are unusable. The simulation results demonstrated here are thus achieved after a full-chip parasitic extraction with sign-off accuracy calculations. Timing analysis is done first to simulate the speed performance of the chip in the different process corners at four different V_{dd}'s. This requires libraries characterized at all those distinct points, something which is not always available. Next, power analysis is done at the just acquired speed points. Running each power analysis at its own speed performance point is crucial to gain insight in the energy-performance trade-off of the microcontrollers, as well as getting a rough estimate about the minimum energy point energy and supply voltage.

Figure 4.13 shows the speed performance from detailed timing analysis of both prototypes. Taking into account the target supply voltage, both prototypes had the same target frequency. As a result, their speed performance is close to identical at the lowest V_{dd}'s. Due to their difference in logic libraries, the speed mismatch becomes larger at higher V_{dd}, especially in extreme process corners. Note that the speed is highly dependent on the process conditions. Almost two orders of magnitude speed difference is observed between the very best and very worst case process conditions. This issue is elaborated on in Chaps. 5 and 6.

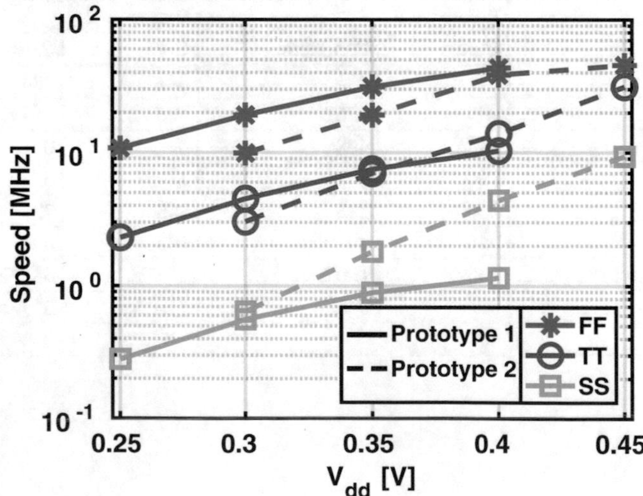

Fig. 4.13 Speed performance from timing analysis of both prototypes at different process corners and supply voltages

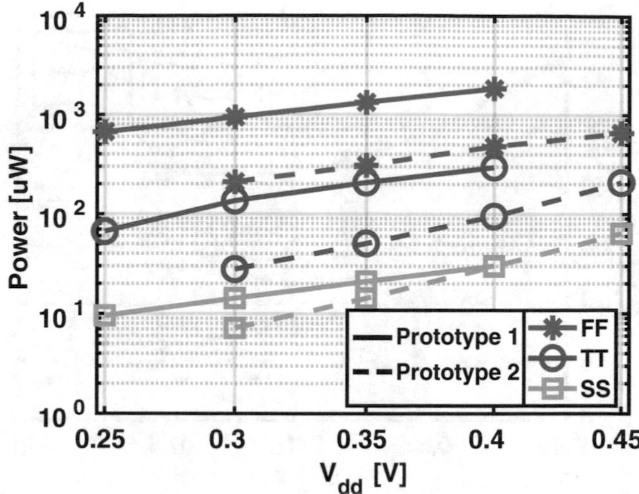

Fig. 4.14 Power consumption from power analysis of both prototypes at different process corners and supply voltages

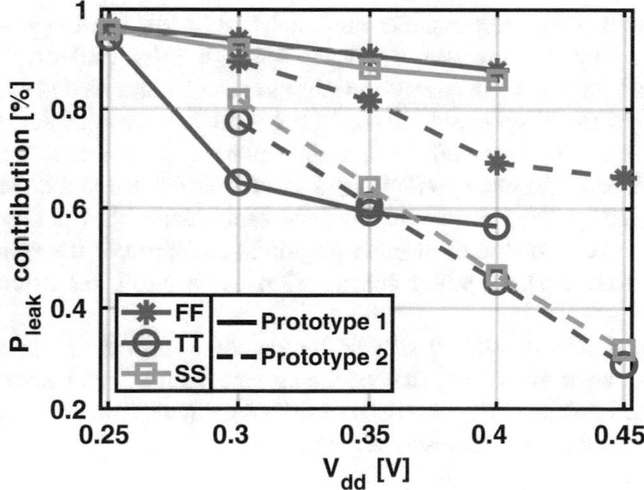

Fig. 4.15 Leakage contribution from power analysis of both prototypes at different process corners and supply voltages

Figure 4.14 shows the power analysis results under the same conditions. A similar effect can be observed under process variations: power consumption shifts two orders of magnitude. Prototype 2 improves significantly on prototype 1: power consumption is reduced by a factor 3–6. The leakage contribution graphs of both prototypes are shown in Fig. 4.15. Finally, Fig. 4.16 shows the energy consumption

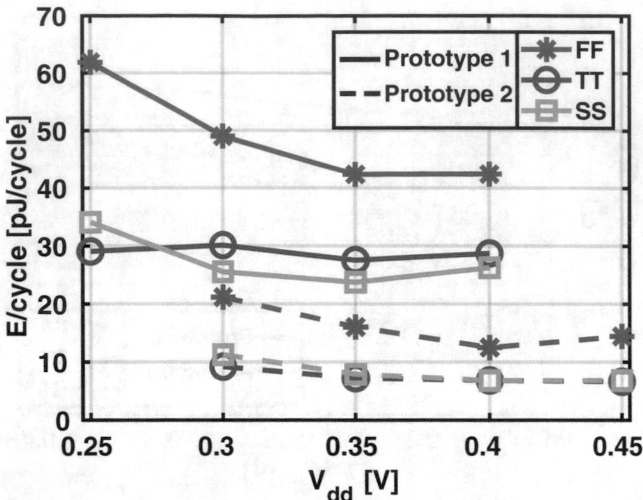

Fig. 4.16 Energy/cycle from power analysis of both prototypes at different process corners and supply voltages

per cycle for both prototypes under all conditions. Again, prototype 2 improves on prototype 1 by lowering energy consumption 3–6 times. Under typical process conditions, a reasonably flat energy consumption can be observed across the entire voltage range. The steep speed incline in Fig. 4.13 balances out the steep power increase in Fig. 4.14. Energy efficiency is thus more or less constant. While process conditions do shift the energy efficiency a bit, the difference is a lot smaller than when considering speed or power only. The same mechanism is observed here: slow (fast) process conditions degrade (improve) the speed of the system as well as decrease (increase) the power consumption. As a result, the overall effect is limited.

Prototype 2 achieves sub-10 pJ/cycle for speeds well in the MHz-range. Such performance can be considered state-of-the-art (see Sect. 4.6) and provides enough motivation to implement the discussed designs in silicon. The observed measurement results are reported and discussed in the next section.

4.4 Measurements

This section shows measurements of the silicon implementations of the microcontroller system presented in Sect. 4.2. The silicon implementation details, the measurement setup and the achieved results are presented. Prototype 1 and 2 implementations are compared displaying how and why prototype 2 improves on prototype 1.

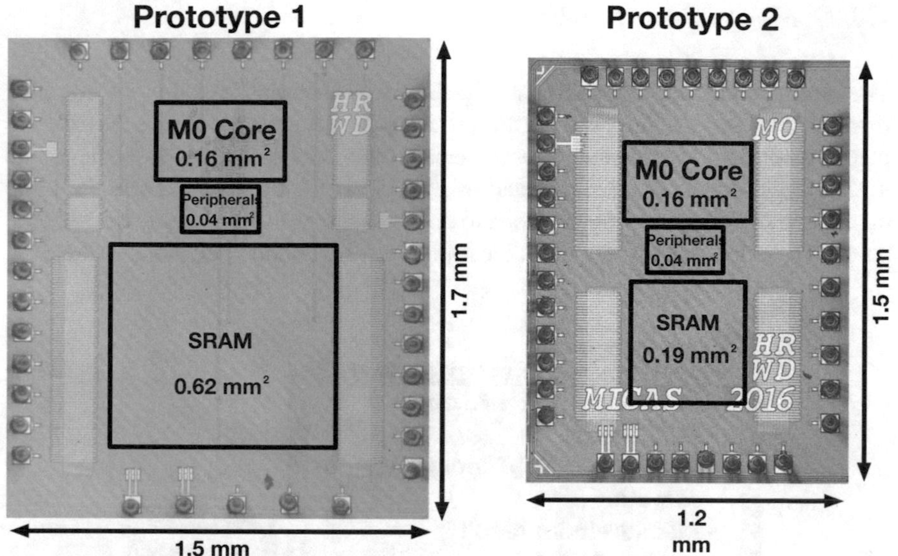

Fig. 4.17 Chip micrograph of both prototypes, fabricated in 40 nm CMOS technology. Left: prototype 1. Right: prototype 2

4.4.1 Silicon

Prototypes 1 and 2 are implemented in 40 nm general-purpose CMOS technology after two separate tape-outs. The chip micrograph of both implementations is shown in Fig. 4.17. The core area of 0.16 mm^2 remained identical between both prototypes. The main reason is they were allocated the same power domain area in both implementations. Their fill factors and density differ slightly. Prototype 2 equips a far smaller SRAM, which is reflected in the total chip area: 2.55 mm^2 for prototype 1 and 1.8 mm^2 for prototype 2. The chips are wire bonded on a PGA package to interface with the measurement setup. The bond pads facilitate power delivery and input–output signalling to control the microcontroller and its peripherals. All measurements reported are arithmetic means of the results acquired from multiple prototype dice. Twenty-five dice of prototype 1 were measured. 8 dice of prototype 2 were measured. The distribution of measurement results by means of a boxplot can provide additional insight in the performance under influence of process variations. The histograms in Figs. 4.24, 4.25, 4.26, 4.27 do the same, as well as the temperature contour plot in Fig. 4.28.

4.4.2 Measurement Setup

The measurement setup presented in Fig. 4.18 was essential in testing the complex functionality and performance of the prototypes. Because the prototypes do not incorporate clock generation, power management and power measurement, the measurement setup takes care of these as well. The PC, the measurement PCB and the measurement equipment cooperate to do semi-autonomous measurements of the prototype-under-test. The PC runs object-oriented JAVA code that controls the PCB

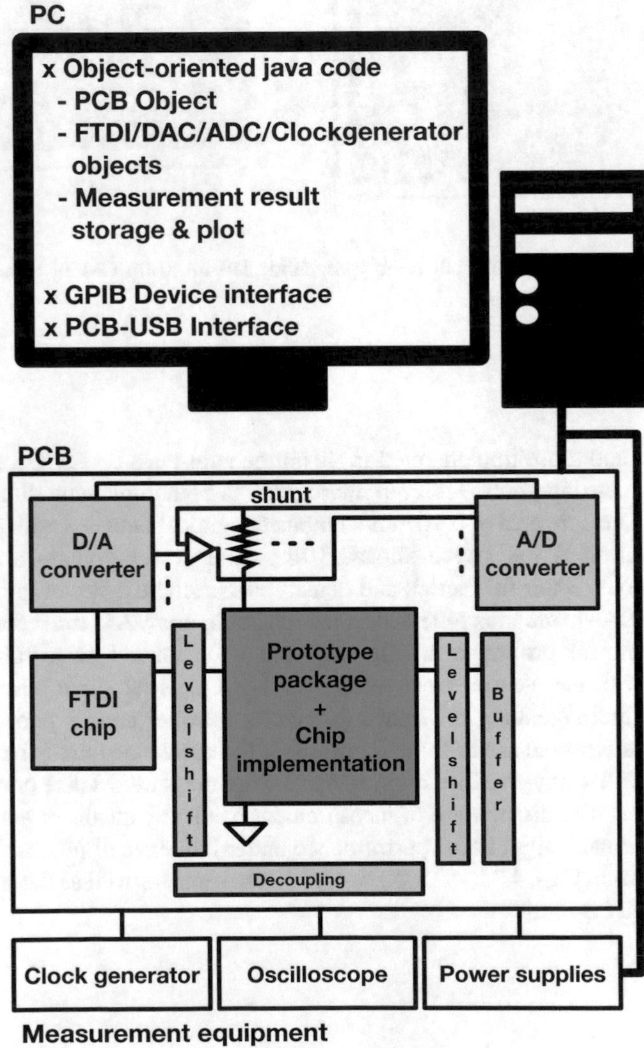

Fig. 4.18 Measurement setup for prototype measurement

and the components on it through an FTDI interface. A D/A converter generates the different supply voltages. An A/D converter is used to measure the current going into the chip using a resistive shunt. A pulse generator generates the clock signal. The clock period and duty cycle are controlled by the PC. The PC runs test software in a three phase cascaded approach:

1. Test basic prototype functionality at a conservative voltage and frequency in the following order: serial debug interface, SRAM memory read and write, GPIO and UART, M0 core with simple C-code and M0 core with complex C-code.
2. Test Dhrystone C-code correct functionality while sweeping voltage and frequency to extract functional operating points.
3. Measure current while continuously running Dhrystone program at operational voltage and frequency points.

The resulting functional operating points and current (power) consumption are reported in the following subsections.

4.4.3 Speed

The M0 microcontroller system operates correctly at clock frequencies ranging from 0.8 MHz to 50 MHz for a 190–500 mV supply voltage (see Fig. 4.19). These speeds are more than sufficient for the applications targeted in Sect. 1.6. Prototype 1 was

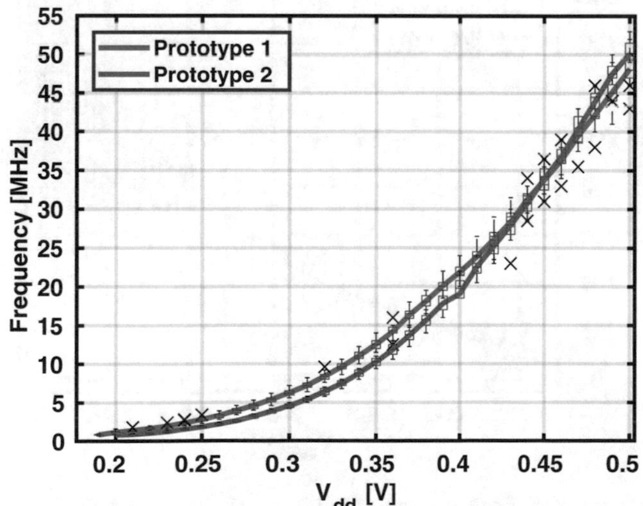

Fig. 4.19 Mean and boxplot of the measured maximum operating frequency as a function of supply voltage for both prototypes

fully functional down to 190 mV. Prototype 2, because of its difference in logic library, functioned down to 200 mV. Although this difference is small, it is non-negligible considering the applied supply voltages. The difference in speed between both prototypes is minimal and can largely be attributed to a slight difference in process corner. The close match in speed shows how the multi-library optimization in prototype 2 is able to achieve identical speed, while still using slower cells for a major part (65.4%) of the design. The boxplots show the variation in speed across different samples. Despite the ultra-low supply voltage resulting in near-threshold operation, speed variation in the form of $\frac{\sigma}{\mu}$ is limited to 8.7% for prototype 1, and 14.7% for prototype 2.

4.4.4 Power and Leakage

While energy consumption is the most important consideration in the systems discussed in this work, one shouldn't lose sight of the absolute power consumption. The applications discussed in Sect. 1.6 can have a variety of power sources, each with their own maximum power output. Power can thus be an important consideration as well. For this reason, Fig. 4.20 shows the measured power of both prototypes operating at the frequencies reported in Fig. 4.19. Power is reduced considerably between prototypes 1 and 2. An average power reduction of 75% is observed for supply voltages up to 300 mV.

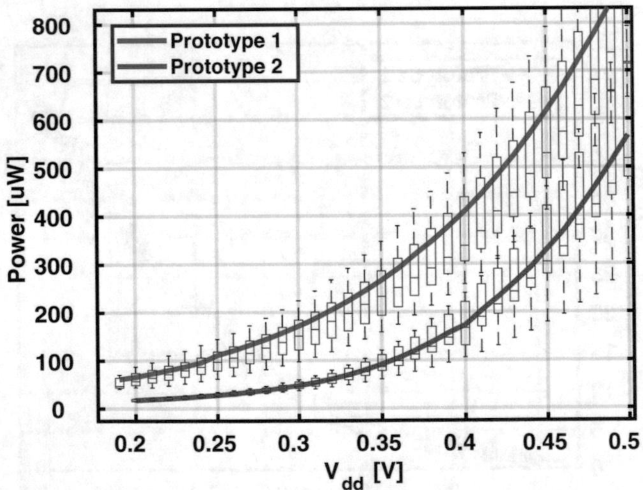

Fig. 4.20 Mean and boxplot of the measured M0 core power as a function of supply voltage for both prototypes at the reported operating speeds

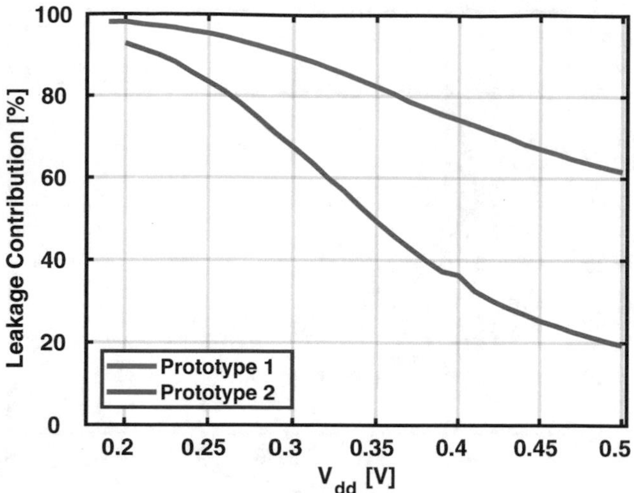

Fig. 4.21 Leakage contribution to M0 core energy consumption as a function of supply voltage for both prototypes

Figure 4.21 shows the relative power that can be attributed to leakage current. The measurements were realized by gating the clock of the circuit and measuring the current consumption. At the lowest supply voltages, leakage power contributes to almost 100% of the total power consumption. With increasing supply voltage, the relative leakage power drops significantly due to the higher switching power and operating frequency. Prototype 2 has significantly lower relative leakage power, influenced by two factors. First, switching power is slightly higher due to the larger input capacitance of the up-sized logic gates with longer gate length. Second, for the same performance and voltage, leakage current is reduced significantly by the dual library optimization and the self-inverting technique. This also has an impact on the minimum energy point, as will be discussed in the next subsection.

4.4.5 Energy

The core energy per cycle of both prototypes is shown in Fig. 4.22. The graphic clearly shows how the lowest V_{dd}'s lead to sub-optimal energy consumption because of the low speed performance and high-leakage power in this V_{dd} range. Prototype 2 significantly improves on prototype 1, in line with the measured speed and power in Figs. 4.19 and 4.20. The average improvement across the entire V_{dd} range is just below 50%. Minimum energy per cycle is achieved at 440 mV, 31 MHz and 16.07 pJ/cycle for prototype 1. Prototype 2 improves on this and achieves a MEP

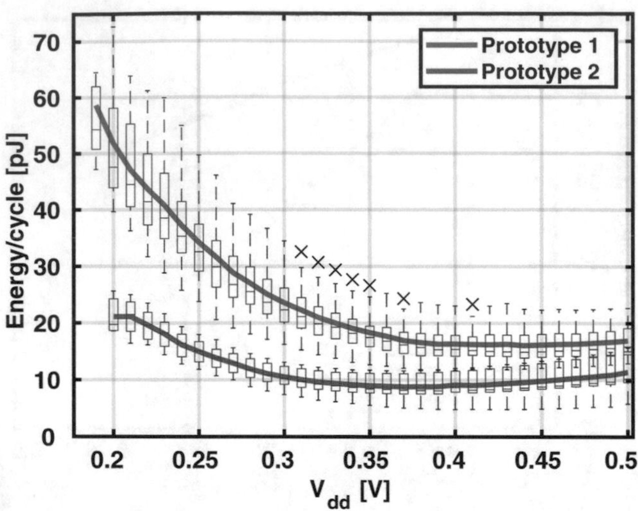

Fig. 4.22 M0 core energy consumption per cycle as a function of supply voltage for both prototypes

of 370 mV, 13.7 MHz and 8.80 pJ/cycle. The MEP of prototype 2 is located at a lower V_{dd}, which means a lower frequency as well. A more detailed energy breakdown of both prototypes is discussed in Sect. 4.5. As both energy curves are quite flat, a wide performance range is possible for energy consumption differing little from that of the MEP. Prototype 2 achieves sub-10 pJ/cycle operation for a frequency range of 6–35 MHz. Although prototype 2 has a lower speed at the MEP, it outperforms prototype 1 across the entire voltage and performance range. This is shown more clearly in Fig. 4.23: a better energy-delay product (EDP) is achieved for all supply voltages. This means the design performs better for all voltage–frequency combinations.

4.4.6 Variations

Considering the ultra-low supply voltage at which the presented systems are being operated, variations can have a big impact on the performance. The boxplots shown in Figs. 4.19, 4.20, 4.21, 4.22 give some insight regarding the speed, power and energy variations between different dice. In the proposed test setup, these variations can largely be attributed to process variations, both inter- and intra-die. Slight temperature and voltage variations may have an impact as well. It was clear from Fig. 4.19 that the relative speed variation is smaller than the relative power variation. The histograms in Figs. 4.24, 4.25, 4.26, 4.27 provide additional insight. Figures 4.24 and 4.25 show how the minimum and MEP supply voltage

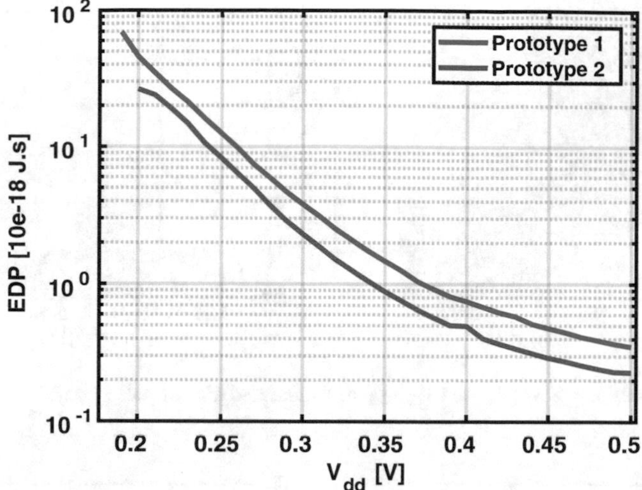

Fig. 4.23 M0 core energy-delay product as a function of supply voltage for both prototypes

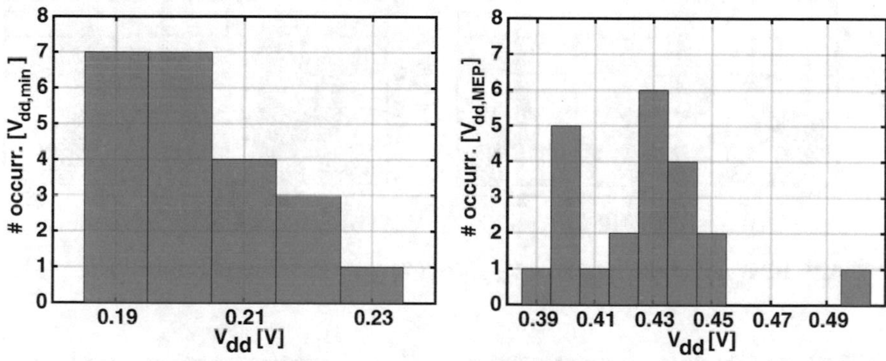

Fig. 4.24 Histogram of $V_{\mathrm{dd,min}}$ and $V_{\mathrm{dd,MEP}}$ of 25 measured dice of prototype 1

vary between different dice. Especially the MEP supply voltage can vary quite a bit (25% or more). Figures 4.26 and 4.27 show the energy at minimum and MEP supply voltage. In line with the voltage variation, the energy varies significantly. As the voltage changes, so does the MEP speed performance. As shown in Sect. 4.3.6, process variations have a significant impact on both speed, power and energy consumption.

To mimic process variations to a larger scale than just measuring multiple single-wafer dice, temperature variations can be useful. Figure 4.28 shows the measured energy/cycle coming from a single die. Multiple voltages and temperatures were applied, always operating at the maximum possible frequency of that operating point. As a result, the contour plot shows how the energy consumption changes

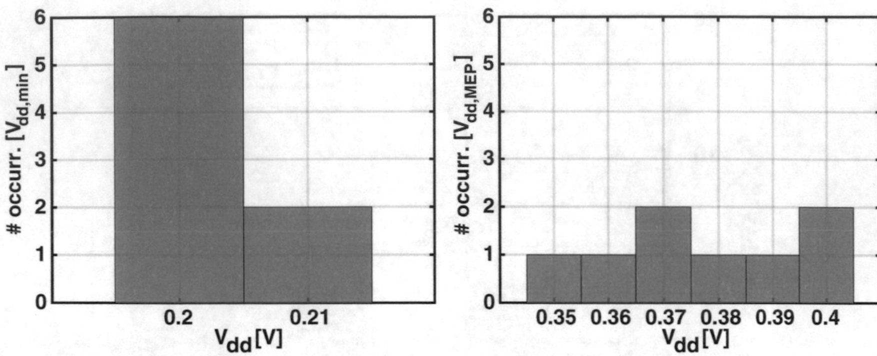

Fig. 4.25 Histogram of $V_{dd,min}$ and $V_{dd,MEP}$ of 8 measured dice of prototype 2

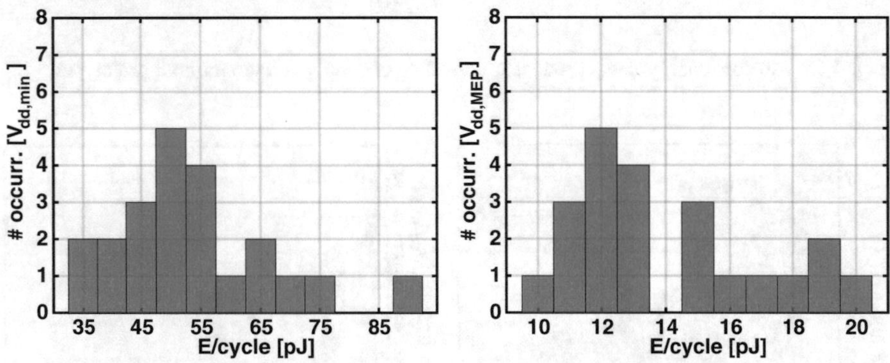

Fig. 4.26 Histogram of energy/cycle at $V_{dd,min}$ and $V_{dd,MEP}$ of 25 measured dice of prototype 1

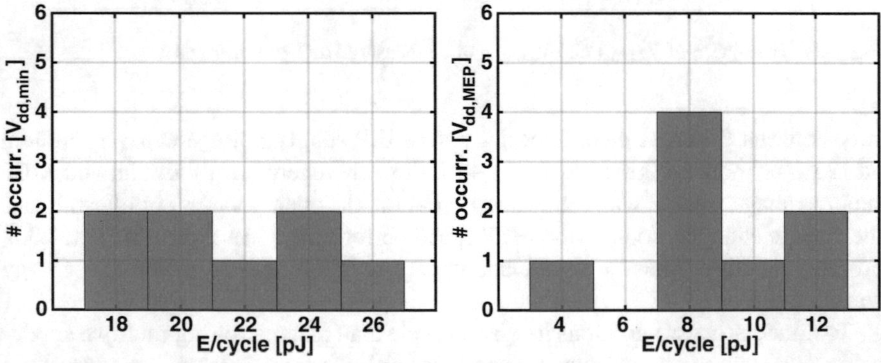

Fig. 4.27 Histogram of energy/cycle at $V_{dd,min}$ and $V_{dd,MEP}$ of 8 measured dice of prototype 2

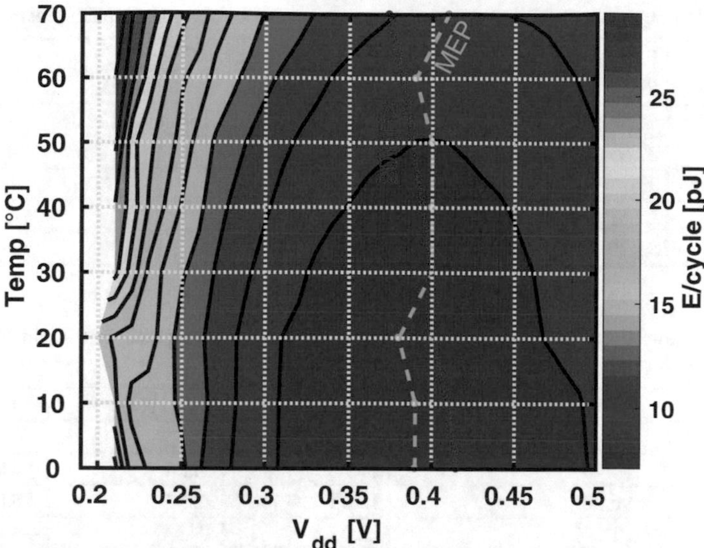

Fig. 4.28 Energy consumption of the M0 core of prototype 2 as a function of supply voltage for a 0–70 °C temperature range

under influence of temperature and voltage. While energy does increase with increasing temperature (+40%), the relative increase is much less then it would be at nominal supply voltage. This is in line with the conclusions of Sect. 2.1.6 regarding temperature sensitivity: for near-threshold operation the effect on operating speed due to temperature variations is limited. The red 20 MHz-line shows that achieving equi-performance in a limited supply voltage range is quite easy. The same holds for the green MEP-line: the MEP voltage changes only slightly when applying temperature variations.

4.4.7 Simulation Mismatch

Table 4.3 compares the achieved performance and energy at 300 mV of the simulated model and the silicon implementation. Both frequency and energy differ significantly in the silicon implementation. The MEP supply voltage and energy also differ significantly. Especially the high MEP supply voltage for prototype 2 is counter intuitive. A decrease in static energy results in a lower MEP supply voltage and energy (see Sect. 1.2). The measurement results in the previous sections clearly showed this.

To further investigate the mismatch between simulation and measurement results, Table 4.4 compares both prototypes with simulation results acquired at identical voltage and frequency conditions. For each voltage, the measured speed and energy

Table 4.3 Comparison of the M0 core performance simulation and measurement results

TT	Prototype 1			Prototype 2		
Condition	Sim.	Meas.	Δ	Sim.	Meas.	Δ
$f_{300\,\text{mV}}$	4.50 MHz	6.20 MHz	+37.8%	3.03 MHz	4.58 MHz	+51.2%
$E_{300\,\text{mV}}$	30.09 pJ	23.57 pJ	−21.7%	9.19 pJ	10.52 pJ	+14.4%
$V_{\text{dd,MEP}}$	350 mV	440 mV	+25.7%	450 mV	370 mV	−17.8%
E_{MEP}	27.51 pJ	16.07 pJ	−41.6%	6.62 pJ	8.80 pJ	+32.9%
E_{stat}	59.1%	68.0%	+9%	28.9%	42.9%	+14%

Table 4.4 Same operating condition comparison of the M0 core performance simulation and measurement results

V_{dd} (mV)	Frequency (MHz)	E_{sim} (pJ)	E_{meas} (pJ)	Δ (%)
Prototype 1 @ TT				
250	2.75	24.32	34.24	+41
300	6.20	15.65	23.57	+51
350	12.48	11.23	18.34	+63
400	21.81	9.17	16.34	+78
Prototype 2 @ TT				
300	4.58	6.79	10.52	+55
350	10.25	5.83	8.99	+54
400	19.06	6.02	9.13	+52
450	33.94	6.47	9.80	+51

are compared to a power simulation of each prototype, simulated at that specific frequency. Note that this would lead to timing violations in a detailed timing analysis. The simulation systematically underestimates the energy consumption. However, the simulation does quite accurately predict the minimum energy point.

The observed mismatch between simulated and measured results has a number of causes:

- **Inaccuracy of the SPICE models:** The SPICE models used for characterization of the timing and power models of the logic are less accurate in the near-threshold region [27].
- **Inaccuracy of the characterization:** The non-linear delay modelling (NLDM) approach used for characterization of the timing and power models is less accurate than novel modelling approaches such as composite current source modelling (CCS) or statistical timing models as discussed in Sect. 3.2.
- **Inter-die process variations:** Inter-die process variations can shift performance and energy consumption significantly. The comparison in Tables 4.3 and 4.4 always assumed typical process conditions. While the manufactured silicon was close to typical–typical, a slight shift can cause a big performance difference. This was clearly demonstrated in Figs. 4.13, 4.14, 4.15, 4.16: the min–max simulation results cover a wide operating range that does include the measured results.

Table 4.5 Full system performance of both prototypes at $V_{dd,min}$ and $V_{dd,MEP}$

	Prototype 1		Prototype 2	
	$V_{dd,min}$	MEP	$V_{dd,min}$	MEP
V_{dd}	190 mV	440 mV	200 mV	370 mV
Frequency	0.8 MHz	31.2 MHz	0.8 MHz	13.7 MHz
Core	60.55 pJ	16.07 pJ	21.09 pJ	8.80 pJ
Peripherals	14.40 pJ	5.62 pJ	5.19 pJ	2.77 pJ
Memory	2476.05 pJ	78.65 pJ	152.73 pJ	31.65 pJ
Full system	2551 pJ	100.34 pJ	179.01 pJ	43.22 pJ

- **Intra-die process variations:** Intra-die process variations can significantly shift performance and energy consumption of ultra-low voltage systems. Although typical performance is statistically most likely, some variation can have a large performance impact. In a lognormal distribution, performance does not average out to the same value as the most likely value. Assuming V_T variation is the most important source of performance variation and is distributed normally, a symmetric V_T-shift in both directions results in a significantly higher (lower) average energy (speed) [7].

4.4.8 Full System

The previous sections have only presented results regarding the M0 core. Apart from the core, the memory is a major contributor to the full system energy consumption. The peripherals equipped in the microcontroller system of this work consume a relatively small portion of the total energy, especially since they are not used when running the Dhrystone C-code. Table 4.5 provides the detailed measurement results of the other building blocks at minimum and MEP supply voltage and frequency. Again, minimum supply voltage operation suffers from high static energy consumption. The MEP operating point (or the wide range around it) provides the optimal operating point. The large total energy difference between prototypes 1 and 2 can be explained by the far smaller SRAM memory in prototype 2. It is clear that the microcontroller could benefit significantly from a small low energy SRAM memory combined with a larger non-volatile memory.

4.5 Energy Optimization

4.5.1 Energy Breakdown

It is clear from the simulation results and silicon measurements that prototype 2 significantly improves on prototype 1. Figure 4.29 shows the breakdown of the core energy consumption in static, dynamic and total energy of both prototypes

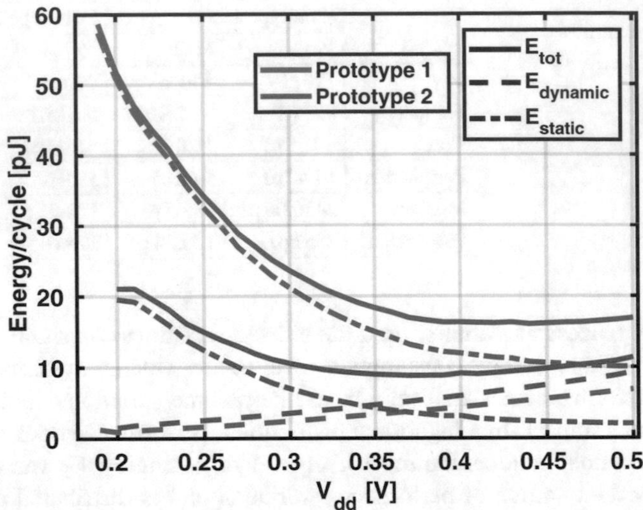

Fig. 4.29 Total, static and dynamic energy trade-off comparison for both prototypes

for the entire voltage range. As discussed earlier, total energy is decreased severely. Looking at Fig. 4.29, this decrease can largely be attributed to a reduction in static energy. Assuming operating frequency is identical for both prototypes, the leakage current of prototype 2 is significantly smaller. The dual logic libraries and leakage optimization techniques resulted in a lower static energy at the same operating frequency.

Comparing dynamic energy, again assuming the same operating frequency is used, a small increase in dynamic energy is observed in prototype 2. The 60 nm gate length logic gates also required a larger gate width to operate with the best possible noise margin at ultra-low voltage. This increase in input capacitance leads to a higher dynamic energy. Both the reduction in static energy and the increase in dynamic energy have the same effect on the minimum energy point: it moves to a lower V_{dd}. As the MEP balances static and dynamic energy, the MEP of prototype 2 can only manifest at a slower speed, voltage and energy. The lower V_{dd} MEP contributes to a dynamic energy reduction despite the higher dynamic energy curve of prototype 2.

4.5.2 f_{max} *vs.* f_{target}

This work did not target a specific application. Therefore, the target frequency was not fixed. All the measurement results presented in this chapter were acquired by operating the circuit at the maximum frequency (f_{max}) for the applied supply voltage. This results in the most optimal operating point for that frequency: both static and dynamic energy are minimized. In designing a system for a specific target frequency, the first consideration should always be the supply voltage. Apart from technology and logic optimizations, the supply voltage is the only variable that can set the target frequency. The second consideration, if possible, should be the logic library. It should be designed to provide reliable operation at the supply voltage attributed to the target frequency while considering process variations. Smaller logic gates result in a smaller dynamic energy consumption for the same supply voltage and operating frequency. The third and last consideration should be the circuit activity. If the application is known, a representative testbench can be used to mimic the correct circuit activity. The dynamic energy consumption is highly related to the activity, shifting the MEP both in voltage and frequency. These three considerations help to set the minimum energy point at or around the target frequency. While a perfect fit between the minimum energy point and the target frequency is desirable, the measurements in this chapter display a flat optimal energy region around the MEP. A slight mismatch between the MEP and the target frequency is not expected to be detrimental to the energy efficiency of the system.

4.5.3 Active vs. Standby Power

In this work, the main focus was to reduce active energy consumption as much as possible, while keeping speed performance reasonably high. In doing so, the effect on leakage power is non-negligible. In applications that duty cycle between data processing and sensing, standby power is of equal importance. Although the leakage optimization of prototype 2 decreases leakage power, it leaves a lot of room for standby power optimization. A finer granularity in power domain division combined with power and clock gating can significantly reduce the leakage power of the entire system in standby mode, as well as sub-blocks of the system that are not being used in active mode. If such power gating is done efficiently, realizing a high-speed performance is often still the best approach: it allows faster data processing and thus a larger relative standby time. The research by Myers et al. [29] is a good example of such an approach. Recent techniques such as dynamic leakage-suppression logic [23] focus on low standby power or minimum overall power instead of minimum energy. The research by Lin et al. [24] is a very recent technique that can switch between both modes with minimal effort. Advanced technologies such as FD-SOI enable body biasing (see Sect. 2.5), which allows relatively easy context switching from low-leakage low-performance to high-leakage high-performance mode and back (Fig. 4.30).

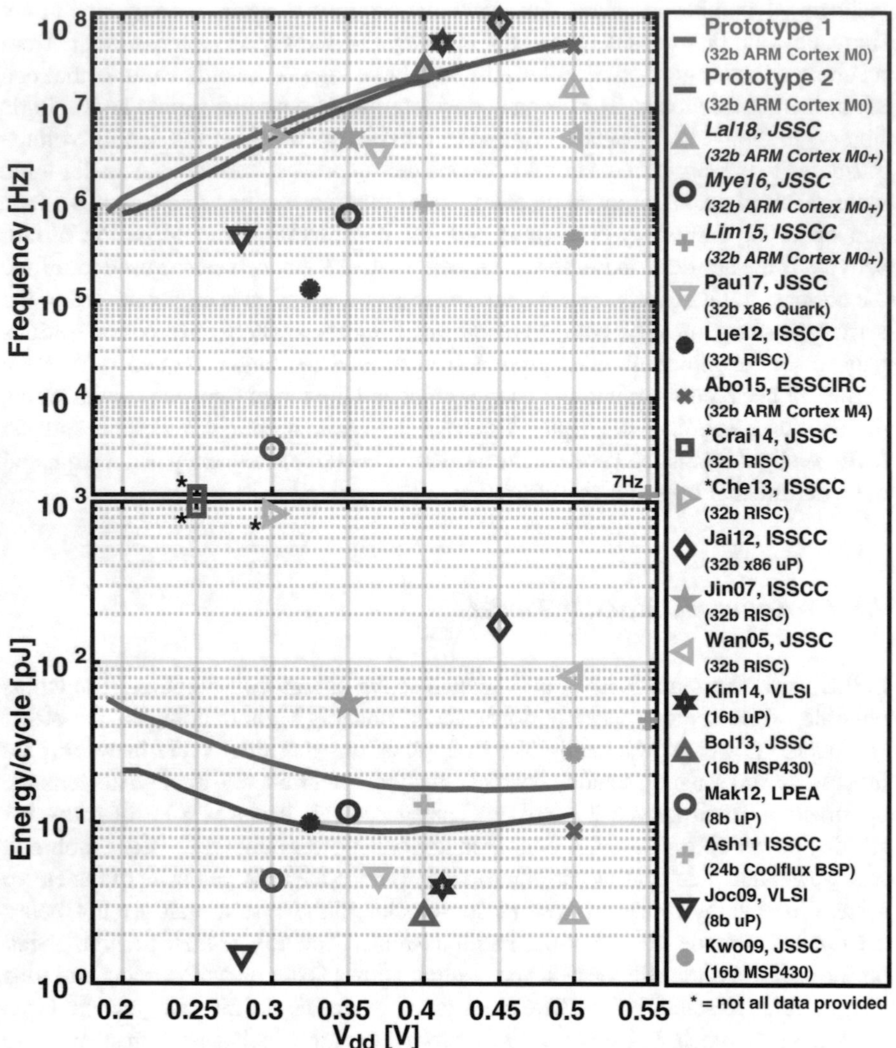

Fig. 4.30 Extensive overview comparing MEP voltage, speed, energy and voltage of the demonstrated prototypes with state-of-the-art ULV microcontroller implementations [1, 5, 6, 9, 11, 15–18, 20, 22, 23, 25, 26, 29, 31, 40]

4.6 State-of-the-Art Comparison

Both prototypes are compared with state-of-the-art 32-bit microcontrollers in Table 4.6. Both very recent publications and slightly older publications are referenced. The referenced designs equip either conventional static CMOS logic [25, 29] or another logic library [23]. Lallement et al. [22] use FD-SOI technology, while

Table 4.6 Performance summary and state-of-the-art comparison

	Prototype 1 [33, 34]	Prototype 2 [34]	JSSC'18, ESSCIRC'17 [21, 22]	JSSC'17 [31]	JSSC16, ISSCC'15 [28, 29]	ISSCC'15 [23]	ISSCC'12 [25]
Technology	40 nm CMOS	40 nm CMOS	28 nm FDSOI	14 nm Tri-Gate CMOS	65 nm CMOS	180 nm CMOS	65 nm CMOS
Core	ARM Cortex M0	ARM Cortex M0	ARM Cortex M0+	32b x86 IA Quark'	ARM Cortex M0+	ARM Cortex M0+	32b RISC
Die area (core)	0.16 mm^2	0.16 mm^2	0.073 mm^2	0.08 mm^2*	0.16 mm^2	1.1 mm^2	0.36 mm^2
Die area (total)	2.55 mm^2	1.80 mm^2	–	0.79 mm^2	3.76 mm^2	2.04 mm^2	2.7 mm^2
Memory	256 KB SRAM @ 0.6 V	64 KB SRAM @ 0.6 V	8 KB SRAM	16 KB ROM, 16 KB CACHE, 64 KB SRAM	2 KB ROM, 8 KB ULV SRAM, 16 KB SRAM	256 B	0.6 KB ULV SRAM, 32 KB SRAM
# dies measured	25	10	5	–	160	28	37
$V_{dd,MEP}$	0.44 V	0.37 V	–	0.37 V	0.35 V	0.55 V	0.325 V
$V_{dd,min}$	0.19 V	0.20 V	0.5 V	0.308 V	0.25 V	0.16 V	0.20 V
Speed @ MEP	31.2 MHz	13.7 MHz	–	3.5 MHz	750 kHz	7 Hz	133 kHz
Speed @ $V_{dd,min}$	0.8 MHz	0.8 MHz	16 MHz	0.5 MHz	27 kHz	16 Hz	10 kHz
Core E/cycle @ MEP	16.07 pJ	8.80 pJ	–	4.64 pJ	/	/	9.94 pJ
Core E/cycle @ $V_{dd,min}$	60.55 pJ	21.09 pJ	0.94 pJ	–			20 pJ*
Total E/cycle @ MEP	100.34 pJ	43.22 pJ	–	17.18 pJ	11.7 pJ	92.04 pJ	/
Total E/cycle @ $V_{dd,min}$	2551 pJ	179.01 pJ	2.67 pJ	45 pJ*	32 pJ	2275 pJ*	
	CPU, Periph, 256 KB	CPU, Periph, 64 KB	CPU, AON, DDSS, Mem	CPU, AON, Mem	CPU, 4 KB SRAM	CPU, 256 B	
EDP [pJ/MHz]	0.5 (core)	0.6 (core)	0.06 (core)	1.3 (core)	/	/	74.4 (core)
	81.2 (total)	3.2 (total)	10.17 (total)	4.9 (total)	15.6 (total)	10^7 (total)	/

*= estimated

[31] use finFET technology. This subset of referenced work allows comparison of the speed and energy performance of the proposed prototypes using transmission gate logic, against conventional static CMOS design as well as more advanced technologies. Lallement et al. [22], Lim et al. [23] and Myers et al. [29] all use a similar architecture. Note that the Cortex-M0+ has been reported to improve area and energy performance by at least 33% compared to the Cortex-M0 [3]. Luetkemeier et al. [25] use a much smaller 32-bit RISC architecture, while [31] use an Intel Quark derived core.

Considering the different architectures and technologies, the ULV optimized differential library of this work does not compromise area nor does the increased gate length used for low-leakage cells as in prototype 2. As expected, the proposed near-threshold libraries result in a significantly higher speed than all other compared designs, even the ones with more advanced technologies. This is a significant benefit over both conventional static CMOS and other logic families or technologies. The single-library design (prototype 1) combines this speed with a low energy consumption, resulting in a very competitive speed–energy combination. The dual library design (prototype 2) improves this even further, resulting in a higher speed and better or comparable core energy consumption than the referenced designs and architectures. Although measurements at the minimum supply voltage confirm the sub-optimal operation in this region, they act as a useful parameter for variation-resilient operation.

When comparing total energy rather than core energy, the memory architecture comes in to play. This work equips a (far) larger memory than most other designs that leaves a lot of room for energy optimization. A memory architecture similar to [29] is expected to emphasize the good speed and energy performance of this work even further.

Finally, the speed–energy combination achieved for both designs is the best metric to compare both speed and energy performance. The research by Lallement et al. [22] is a very recent work in 28nm FDSOI and is the only one outperforming prototype 2. All others have an energy-delay product at least 5 times larger. Figure 4.31 shows how the proposed prototypes and the most comparable state-of-the-art 32-bit implementations trade off speed and energy consumption.

A more extensive overview of the literature is shown in Fig. 4.30, where the MEP voltage, speed and energy consumption are compared to the demonstrated performance of this work. As visible in the upper half of the figure, the proposed prototypes outperform almost all designs in speed, no matter the architecture. The bottom half of the figure shows how the proposed prototypes perform similar to other work when considering energy/cycle, especially when considering the architecture and its bit width. When any of the most recent referenced works improve on speed or energy, it always comes at the cost of the other.

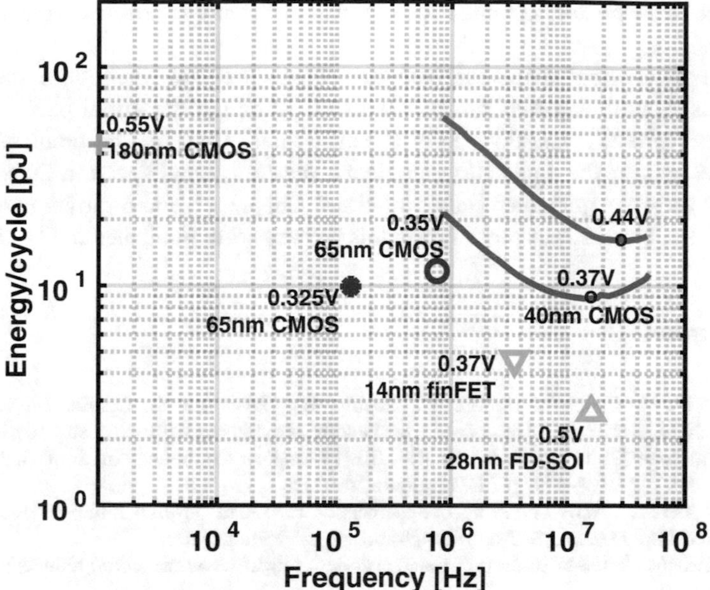

Fig. 4.31 Literature overview comparing the speed–energy combination of the demonstrated prototypes with other state-of-the-art prototypes that have similar 32-bit architectures

4.7 Conclusion

This chapter discussed the application, design, implementation and silicon measurements of two generations of voltage-scaled microcontroller systems. These systems consist of an ARM Cortex-M0 core, an SRAM memory and the necessary peripherals for testing and debugging. The differential transmission gate design flow presented in Chap. 3 is used to implement two prototypes in 40 nm CMOS. The prototypes are fully functional and compatible with the ARM tool chain and C-code compiler framework.

Scaling of the supply voltage of the described systems allows minimum energy operation. The energy breakdown of both prototypes provides insight in the energy consumption and demonstrates how and why prototype 2 improves on prototype 1, resulting in a lower supply voltage and energy consumption minimum energy point. Both prototypes have proven to be competitive regarding energy consumption, and outperform state-of-the-art when considering speed performance. As such, they enable intensive processing applications at low voltage, compatible with the typical IoT-node power footprint.

The energy consumption of the memory system used to run the M0 core leaves room for improvement. This work focused on achieving a relatively high-speed performance combined with low energy consumption. A variety of supplementary techniques can be implemented to improve the standby power of the proposed system.

When simulation results of both prototypes are compared with their measurements, some mismatch can be observed. Although measured performance is very good, process variations have a detrimental effect on the predictability of the prototypes. If the application requires a predetermined speed, a DVS-enabled feedback loop can guarantee such conditions. Chapters 5 and 6 further explore this strategy. The conclusions presented in this chapter were published in [14, 38, 39].

References

1. Abouzeid, F., Clerc, S., Bottoni, C., Coeffic, B., Daveau, J.M., Croain, D., Gasiot, G., Soussan, D., Roche, P.: 28 nm FD-SOI technology and design platform for sub-10 pJ/cycle and SER-immune 32bits processors. In: 41st IEEE European Conference on Solid-State Circuits (ESSCIRC), pp. 108–111. IEEE, Piscataway (2015)
2. ARM: AMBA 3 AHB-Lite Protocol Specification v1.0 Arm Developer. https://developer.arm.com/docs/ihi0033/a/amba-3-ahb-lite-protocol-specification-v10
3. ARM: ARM Cortex-M series processors overview. https://www.arm.com/products/processors/cortex-m
4. ARM: DesignStart for University arm. https://www.arm.com/resources/designstart/designstart-university
5. Ashouei, M., Hulzink, J., Konijnenburg, M., Zhou, J., Duarte, F., Breeschoten, A., Huisken, J., Stuyt, J., de Groot, H., Barat, F., David, J., Van Ginderdeuren, J.: A voltage-scalable biomedical signal processor running ECG using 13 pJ/cycle at 1 MHz and 0.4 V. In: IEEE International Solid-State Circuits Conference Digest of Technical Papers (ISSCC), pp. 332–334. IEEE, Piscataway (2011)
6. Bol, D., De Vos, J., Hocquet, C., Botman, F., Durvaux, F., Boyd, S., Flandre, D., Legat, J.D.: SleepWalker: a 25-MHz 0.4-V Sub-mm2 7-uW/MHz Microcontroller in 65-nm LP/GP CMOS for Low-Carbon Wireless Sensor Nodes. IEEE J. Solid State Circuits **48**(1), 20–32 (2013)
7. Bowman, K., Duvall, S., Meindl, J.: Impact of die-to-die and within-die parameter fluctuations on the maximum clock frequency distribution for gigascale integration. IEEE J. Solid State Circuits **37**(2), 183–190 (2002)
8. Chen, W., Hancke, G., Mayes, K., Lien, Y., Chiu, J.H.: NFC mobile transactions and authentication based on GSM network. In: Second International Workshop on Near Field Communication, pp. 83–89. IEEE, Piscataway (2010)
9. Chen, J.S., Yeh, C., Wang, J.S.: Self-super-cutoff power gating with state retention on a 0.3 V 0.29 fJ/cycle/gate 32 b RISC core in 0.13 um CMOS. In: IEEE International Solid-State Circuits Conference Digest of Technical Papers (ISSCC), pp. 426–427. IEEE, Piscataway (2013)
10. Cobham Gaisler: Leon/GRLIB. https://www.gaisler.com/
11. Craig, K., Shakhsheer, Y., Arrabi, S., Khanna, S., Lach, J., Calhoun, B.H.: A 32 b 90 nm processor implementing panoptic dvs achieving energy efficient operation from sub-threshold to high performance. IEEE J. Solid-State Circuits **49**(2), 545–552 (2014)
12. Flynn, D., Wood, T., Dworsky, P., Melikyan, V., Babayan, E.: Teaching IC design with the ARM Cortex-M0 designstart processor and synopsys 90 nm educational design kit. In: 3rd Interdisciplinary Engineering Design Education Conference, pp. 36–38. IEEE, Piscataway (2013)
13. Hamacher, V., Kornagel, U., Lotter, T., Puder, H.: Binaural signal processing in hearing aids: technologies and algorithms. In: Advances in Digital Speech Transmission, pp. 401–429. Wiley, Chichester (2008)

14. IFixit, Oculus-VR: Oculus Rift CV1 Teardown - iFixit. https://www.ifixit.com/Teardown/Oculus+Rift+CV1+Teardown/60612 (2016)
15. Jain, S., Khare, S., Yada, S., Ambili, V., Salihundam, P., Ramani, S., Muthukumar, S., Srinivasan, M., Kumar, A., Gb, S.K., Ramanarayanan, R., Erraguntla, V., Howard, J., Vangal, S., Dighe, S., Ruhl, G., Aseron, P., Wilson, H., Borkar, N., De, V., Borkar, S.: A 280 mV-to-1.2 V wide-operating-range IA-32 processor in 32 nm CMOS. In: IEEE International Solid-State Circuits Conference Digest of Technical Papers (ISSCC), pp. 66–68. IEEE, Piscataway (2012)
16. Jin, W., Kim, S., He, W., Mao, Z., Seok, M.: In situ error detection techniques in ultralow voltage pipelines: analysis and optimizations. IEEE Trans. Very Large Scale Integr. VLSI Syst. **25**(3), 1032–1043 (2017)
17. Jock, S.C., Bolus, J.F., Wooters, S.N., Jurik, A.D., Weaver, A.C., Blalock, T.N., Calhoun, B.H.: A 2.6-uW sub-threshold mixed-signal ECG SoC. In: IEEE Symposium on VLSI Circuits (VLSI) (2009)
18. Kim, S., Mingoo Seok: R-processor: 0.4 V resilient processor with a voltage-scalable and low-overhead in-situ error detection and correction technique in 65 nm CMOS. In: Symposium on VLSI Circuits Digest of Technical Papers (VLSI), pp. 1–2. IEEE, Piscataway (2014)
19. Kim, H., Kim, S., Van Helleputte, N., Artes, A., Konijnenburg, M., Huisken, J., Van Hoof, C., Yazicioglu, R.F.: A configurable and low-power mixed signal SoC for portable ECG monitoring applications. IEEE Trans. Biomed. Circuits Syst. **8**(2), 257–267 (2014)
20. Kwong, J., Ramadass, Y.K., Verma, N., Chandrakasan, A.P.: A 65 nm Sub-V_t microcontroller with integrated SRAM and switched capacitor DC-DC converter. IEEE J. Solid State Circuits **44**(1), 115–126 (2009)
21. Lallement, G., Abouzeid, F., Cochet, M., Daveau, J.M., Roche, P., Autran, J.L.: A 2.7 pJ/cycle 16 MHz SoC with 4.3 nW power-off ARM Cortex-M0+ core in 28 nm FD-SOI. In: 43rd IEEE European Solid-State Circuits Conference (ESSCIRC), pp. 153–162. IEEE, Piscataway (2017)
22. Lallement, G., Abouzeid, F., Cochet, M., Daveau, J.M., Roche, P., Autran, J.L.: A 2.7 pJ/cycle 16 MHz, 0.7 μW deep sleep power ARM Cortex-M0+core SoC in 28 nm FD-SOI. IEEE J. Solid State Circuits 53(7), 2088–2100 (2018)
23. Lim, W., Lee, I., Sylvester, D., Blaauw, D.: Batteryless Sub-nW cortex-M0+ processor with dynamic leakage-suppression logic. In: IEEE International Solid-State Circuits Conference Digest of Technical Papers (ISSCC), pp. 1–3. IEEE, Piscataway (2015)
24. Lin, L., Jain, S., Alioto, M.: A 595 pW 14 pJ/Cycle microcontroller with dual-mode standard cells and self-startup for battery-indifferent distributed sensing. In: IEEE International Solid-State Circuits Conference Digest of Technical Papers (ISSCC), pp. 44–46. IEEE, Piscataway (2018)
25. Luetkemeier, S., Jungeblut, T., Porrmann, M., Rueckert, U.: A 200 mV 32b subthreshold processor with adaptive supply voltage control. In: IEEE International Solid-State Circuits Conference Digest of Technical Papers (ISSCC), pp. 484–486. IEEE, Piscataway (2012)
26. Mäkipää, J., Turnquist, M.J., Laulainen, E., Koskinen, L.: Timing-error detection design considerations in subthreshold: an 8-bit microprocessor in 65 nm CMOS. J. Low Power Electr. Appl. **2**(2), 180–196 (2012)
27. Markovic, D., Wang, C., Alarcon, L., Liu, T.T, Rabaey, J.: Ultralow-power design in near-threshold region. Proc. IEEE **98**(2), 237–252 (2010)
28. Myers, J., Savanth, A., Howard, D., Gaddh, R., Prabhat, P., Flynn, D.: An 80 nW retention 11.7 pJ/cycle active subthreshold ARM Cortex-M0+ subsystem in 65 nm CMOS for WSN applications. In: IEEE International Solid-State Circuits Conference Digest of Technical Papers (ISSCC), pp. 1–3. IEEE, Piscataway (2015)
29. Myers, J., Savanth, A., Gaddh, R., Howard, D., Prabhat, P., Flynn, D.: A subthreshold ARM Cortex-M0+ subsystem in 65 nm CMOS for WSN applications with 14 power domains, 10T SRAM, and integrated voltage regulator. IEEE J. Solid State Circuits **51**(1), 31–44 (2016)
30. Nachman, L., Huang, J., Shahabdeen, J., Adler, R., Kling, R.: IMOTE2: serious computation at the edge. In: International Conference on Wireless Communications and Mobile Computing, pp. 1118–1123. IEEE, Piscataway (2008)

31. Paul, S., Honkote, V., Kim, R.G., Majumder, T., Aseron, P.A., Grossnickle, V., Sankman, R., Mallik, D., Wang, T., Vangal, S., Tschanz, J.W., De, V.: A sub-cm^3 energy-harvesting stacked wireless sensor node featuring a near-threshold voltage IA-32 microcontroller in 14-nm tri-gate CMOS for always-ON always-sensing applications. IEEE J. Solid State Circuits **52**(4), 961–971 (2017)
32. Pestana, J., Sanchez-Lopez, J.L., Saripalli, S., Campoy, P.: Computer vision based general object following for GPS-denied multirotor unmanned vehicles. In: American Control Conference, pp. 1886–1891. IEEE, Piscataway (2014)
33. Reyserhove, H., Dehaene, W.: A 16.07pJ/cycle 31 MHz fully differential transmission gate logic ARM cortex M0 core in 40 nm CMOS. In: 42nd IEEE European Conference on Solid-State Circuits (ESSCIRC), pp. 257–260. IEEE, Piscataway (2016)
34. Reyserhove, H., Dehaene, W.: A Differential transmission gate design flow for minimum energy sub-10-pJ/Cycle ARM Cortex-M0 MCUs. IEEE J. Solid State Circuits **52**(7), 1904–1914 (2017)
35. Rooseleer, B., Cosemans, S., Dehaene, W.: A 65 nm, 850 MHz, 256 kbit, 4.3 pJ/access, ultra low leakage power memory using dynamic cell stability and a dual swing data link. IEEE J. Solid State Circuits **47**(7), 1784–1796 (2012)
36. Seok, M., Blaauw, D., Sylvester, D.: Clock network design for ultra-low power applications. In: Proceedings of the 16th ACM/IEEE International Symposium on Low Power Electronics and Design (ISLPED), p. 271. IEEE, Piscataway (2010)
37. Sharma, V., Cosemans, S., Ashouie, M., Huisken, J., Catthoor, F., Dehaene, W.: Ultra low-energy SRAM design for smart ubiquitous sensors. IEEE Micro **32**(5), 10–24 (2012)
38. Sparkfun, NEST: Nest Thermostat Teardown. https://learn.sparkfun.com/tutorials/nest-thermostat-teardown- (2016)
39. TechInsights, Fitbit: Fitbit Charge 2 Teardown. http://www.techinsights.com/about-techinsights/overview/blog/fitbit-charge-2-teardown/
40. Wang, J.S., Chen, J.S., Wang, Y.M., Yeh, C.: A 230 mV-to-500 mV 375 KHz-to-16 MHz 32b RISC Core in 0.18 μm CMOS. In: IEEE International Solid-State Circuits Conference Digest of Technical Papers (ISSCC), pp. 294–604. IEEE, Piscataway (2007)
41. Warneke, B., Pister, K.: An ultra-low energy microcontroller for Smart Dust wireless sensor networks. In: IEEE International Solid-State Circuits Conference Digest of Technical Papers (ISSCC), pp. 316–317. IEEE, Piscataway (2004)
42. Weicker, R.P.: Dhrystone benchmark (version 2.1) http://groups.google.com/group/comp.arch/browse_thread/thread/b285e89dfc1881d3/068 (1988)
43. Yiu, J.: The definitive guide to the ARM Cortex-M0 (2011)

Chapter 5
Error Detection and Correction

Abstract In-field operating conditions of an integrated circuit can vary significantly. Both manufacturing (process corners and mismatch) and non-manufacturing (temperature, voltage, ageing, etc.) induced factors can influence the operation of a circuit. Especially when operating in the near-threshold region, circuits are highly susceptible to these variations. Chapter 4 clearly demonstrated this: while the measured prototypes show ultra-low energy consumption at MHz-speed performance, measuring multiple dice at multiple temperatures shows variations in the achieved results. Moreover, the simulation results showed a large span in frequency and power consumption under process variations.

Traditional solutions to overcome this unpredictability in operating frequency have been to over-design the circuit, as to meet design specifications even under worst case conditions. In digital logic circuits, as with most circuits, this results in an overhead. The circuit is over-designed to make the slowest pipeline stage meet the target operating frequency under the worst conditions. The consequences are a reduced maximum clock frequency and/or increased total energy consumption. Under nominal conditions, the circuit would be able to operate either faster or at a lower supply voltage. While real-time conditions change (e.g., temperature), process variation impact is fixed after fabrication. Operating a circuit fabricated under typical conditions as if it were fabricated under the worst conditions is a very naive way to guarantee performance.

For near-threshold operation, the worst case margined approach is detrimental to the ultra-low energy consumption and high-speed operation it aims to achieve. Timing error detection and correction (EDAC) is a technique to adaptively overcome most of these safety margins under varying conditions. A lot of work has been presented to realize better-than-worst-case performance through EDAC at (close to) nominal supply voltages. However, few works enable such techniques for ultra-low voltage operation, while these circuits can benefit the most. By enabling such operation, near-threshold circuits can be leveraged to their full potential. This chapter goes into further detail on better-than-worst-case design techniques and error detection and correction systems. It discusses several of the most important

H. Reyserhove, W. Dehaene, *Efficient Design of Variation-Resilient Ultra-Low Energy Digital Processors*, https://doi.org/10.1007/978-3-030-12485-4_5

works and navigates the different design trade-offs to be made in such a system. As such, it lays out the ground work for the EDAC-enabled microcontroller discussed in Chap. 6.

Section 5.1 discusses the different strategies to improve circuit predictability. An overview of a typical EDAC system is given in Sect. 5.2, setting the framework for further considerations. Section 5.3 looks at other (historical) margin reduction techniques, giving context to error detection and correction strategies. Section 5.4 introduces the different EDAC concepts and discusses them in depth. Section 5.5 considers the aspects particular to ultra-low voltage and variation-sensitive operation. An EDAC system should improve the baseline circuit it comprises. In Sect. 5.6, considerations on how to quantify this improvement are presented. Finally, Sect. 5.7 draws a conclusion.

5.1 Predictability-Enabling Strategies

5.1.1 Silicon Lottery

Advanced nanometer CMOS technologies inevitably suffer from higher relative variations. While production techniques continue to advance, these technologies require atomic-level precise manufacturing. The efforts discussed in Chap. 2 do manage variability to some extent, but the combination of nanometer scale CMOS and near-threshold logic inevitably leads to bad performance predictability. Producing ICs in an advanced CMOS technology in high volume could be considered as playing the 'silicon lottery'. Every manufactured die can have a range of imperfections leading to different performance characteristics. When considering the mismatch between different dice, their performance can be displayed using a probability density function as shown in Fig. 5.1.

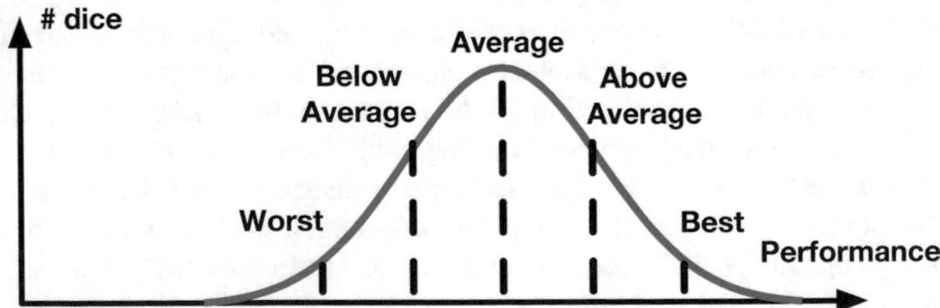

Fig. 5.1 Conceptual performance qualification probability density function of a large distribution of dice

5.1.2 Worst Case Strategy

In the ideal case every die is able to operate at its fabricated performance, meaning no design margin is equipped. However, in selecting a random die from this distribution, no better than a worst case based performance can be expected (see Fig. 5.2). As such, normal clock edge triggered logic is designed accordingly, setting design specifications so that the worst case performance is always met. The consequence is a huge design margin for most of the dice. At the same time, the average die, having the highest occurrence in the distribution, is able to perform far better than the worst case.

5.1.3 Binning

One solution to overcome a large part of the design margin is binning. By sorting the manufactured dice in three or more bins according to measured performance, a smaller design margin can be equipped for each bin, as shown in Fig. 5.3. If the application allows it, different grade products can become available in this way. Although this strategy is effective, it requires post-fabrication performance characterization of each die. This usually means testing the die under a range of voltage–frequency ambient conditions to determine its manufactured performance.

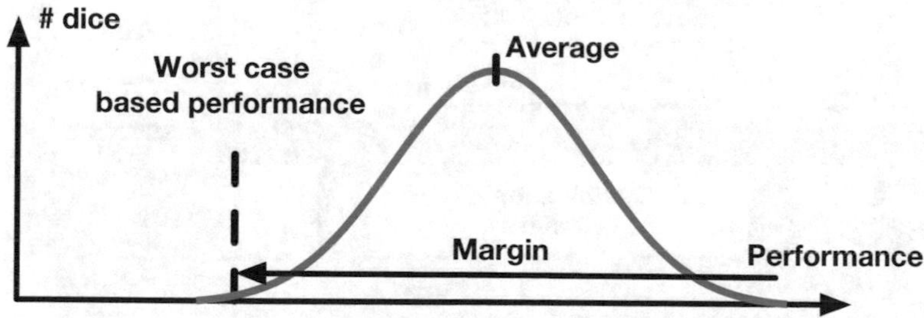

Fig. 5.2 Worst case based strategy to enable predictable performance

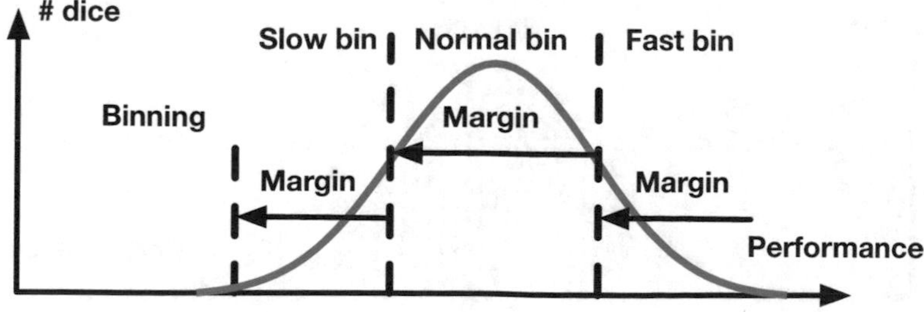

Fig. 5.3 Binning strategy to enable predictable performance

The time and cost of such a test strategy is non-trivial. Additionally, depending on the amount of bins, the margins needed for each bin are still rather large.

5.1.4 Replica Monitoring

A better and often applied approach to overcome the poor performance predictability is the use of replica circuitry. A replica of the circuits critical path (often referred to as a canary circuit) is equipped on the same die and its performance is monitored continuously. Figure 5.4 shows a conceptual block diagram of this approach as applied in [25]. The replica path is used as a reference to predict actual circuit performance. The performance can then be tuned using voltage and/or frequency scaling. The replica path shares process corner, global voltage and global temperature with the actual critical path, thus predicting its performance way better than design time based prediction like in Fig. 5.2. By slightly under-designing the replica, the monitor makes sure the replica fails before the actual critical path, providing an always-correct monitoring scheme. This is achieved by adding margin to the replica path as shown in Fig. 5.4. Figure 5.5 demonstrates that this results in a continuous performance range which equips a smaller margin at every performance

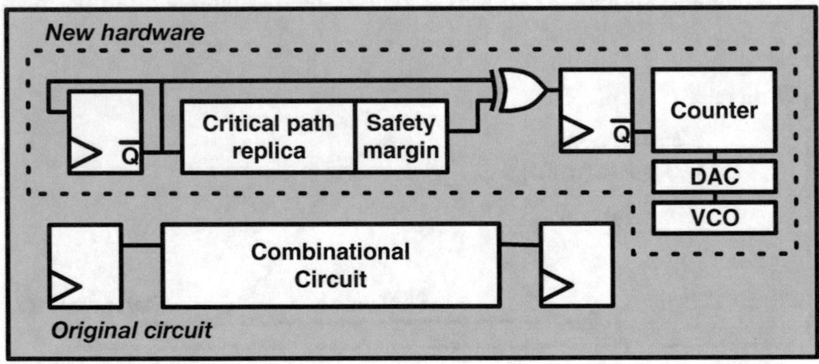

Fig. 5.4 Example replica circuit implementation as shown in [25]

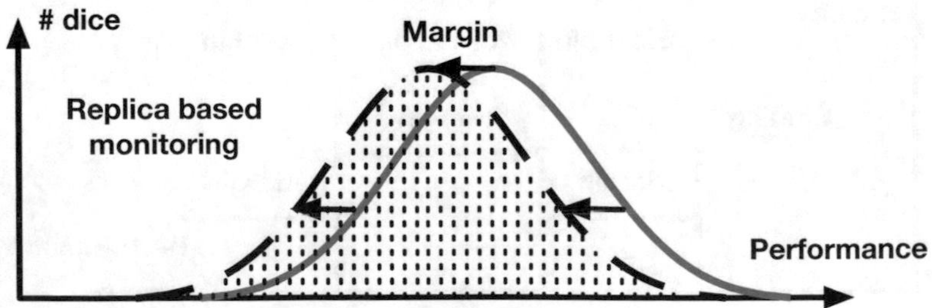

Fig. 5.5 Replica circuit strategy for performance qualification

point. The replica cannot track intra-die variations resulting in mismatch between the replica and the actual critical path, as well as local fluctuations in temperature, supply voltage and ageing. Furthermore, mismatch between the replica and the actual critical path can vary with voltage/frequency scaling. Although better than binning or worst case performance qualification, the replica strategy still results in significantly large margins due to on-chip variation and is detrimental to the overall system performance when it is applied for ultra-low voltage operation.

5.1.5 In Situ Monitoring

In situ monitoring can overcome the mismatch issue replica monitoring faces. By monitoring the behaviour of the actual circuit rather than a replica, the influence of on-chip variation can be eliminated. When it comes to predicting maximum performance, there is no better alternative than monitoring the circuit itself. Margins can be reduced fully as local fluctuations in temperature or supply voltage are equally present in the monitor (see Fig. 5.6).

Operating without margins through in situ monitoring leads to near-failure execution. In changing ambient conditions this point-of-failure is a moving target. Dynamic voltage or frequency scaling is typically used to change the circuit performance accordingly. While in situ monitoring aims to successfully track the point-of-failure, instantaneous tracking is to optimistic. Adaptive voltage scaling requires time to set a new target voltage in the power management unit and settle the voltage across the circuit. Moreover, some ambient conditions change instantaneously. Faulty execution becomes inevitable. This introduces logic errors in the circuit which results in system malfunction. Error detection and correction (EDAC) circuits can be inserted to detect and overcome these errors.

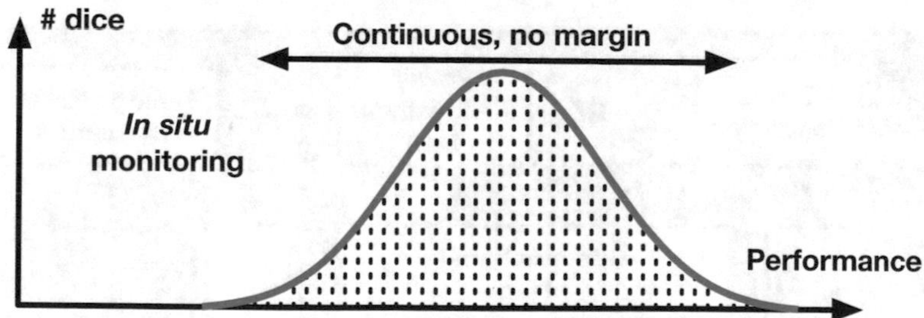

Fig. 5.6 In situ monitoring performance

5.2 System Overview

Before discussing different error detection and correction techniques, this section sketches a typical error detection and correction enabled system. In the context of sequential logic, errors are almost always timing errors: data-paths requiring more time to propagate than provided by the clock period, violating the setup time of the capturing flip-flop. Figure 5.7 shows how a typical EDAC system is composed of two loops. In the first loop, the original circuit is augmented with an error detection system. An error correction system uses the error detection information to overcome errors and guarantee forward progress of the pipelined circuit. The second loop uses the error information gathered through error detection to control either the supply voltage, the clock frequency or both.

The typical mode of operation would be to aggressively scale the supply or the frequency until error detection displays errors. This point is called the point-of-first-failure (PoFF). Operating at this point, voltage and/or frequency margins are effectively eliminated. In the meanwhile the error correction system continues to guarantee correct operation. By continuously monitoring error rates and adjusting the voltage or frequency, the PoFF is tracked despite (fast) changing ambient conditions. Additionally, in a digital circuit the critical path is normally the first to fail. The point-of-first-failure thus tracks the critical path similar to replica circuit based monitoring. The big difference however is that this PoFF strategy can adaptively monitor any critical path under any condition continuously. This is particularly convenient in conditions where it is hard to predict critical paths and their performance due to high variation sensitivity.

Most EDAC systems operate in the way just described. They differentiate in error detection and correction strategies and circuits. In designing an EDAC system, the goal should always be to improve overall performance, e.g., minimize total energy consumption. A complex system as depicted in Fig. 5.7 introduces hardware and

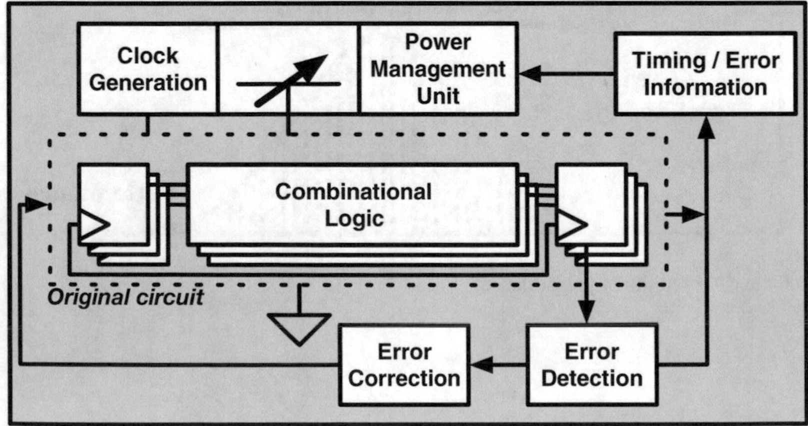

Fig. 5.7 System overview of a typical EDAC system

energy overhead on different levels. Since margins can be reduced significantly, a performance increase (V_{dd} reduction or frequency increase) is evident. The key in designing an EDAC system is to balance these two effects to reduce overall energy, ideally while improving the original system by a non-trivial amount.

5.3 History

One of the first concepts that inherently overcomes the need for margining is completion detection. While this technique is asynchronous in nature and thus does not benefit from the same advantages as clock edge triggered sequential logic, it is often enabled to provide self-timing [20]. In a sense, this is the truest form of timing information on data-path signals. Since the information on timing is similar to EDAC strategies, it is of value for the purpose of this study.

Monitoring the transient current flow of a circuit is a convenient analog way to detect completion of a given logic function. Figure 5.8 shows an example of such current-sensing completion detection (CSCD) [1, 10]. The Muller C-element [6, 14] (see Fig. 5.9) is often referred to combinatorial way of handling completion

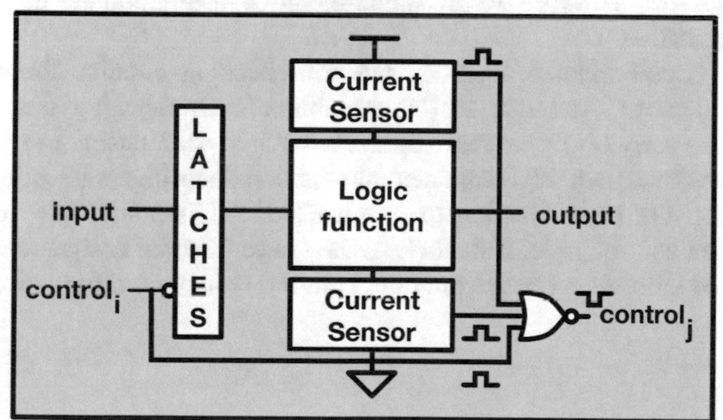

Fig. 5.8 Current-sensing completion detection (CSCD) as proposed in [10]

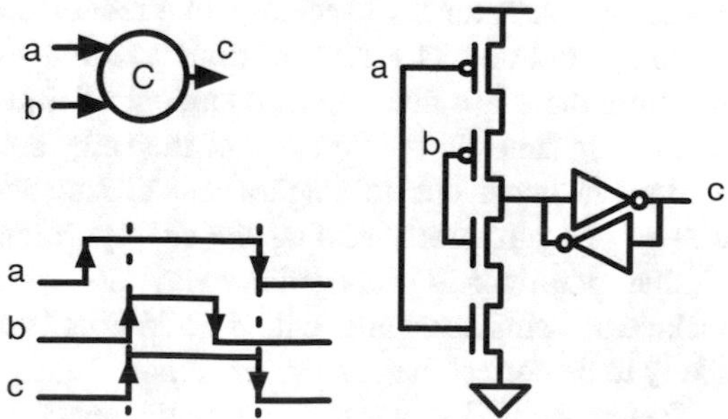

Fig. 5.9 Muller C-element

detection. It can combine the acknowledge signals typically used in asynchronous logic. Analog completion detection strategies like in [10] are often preferred because of their area efficiency and their transparency for the monitored logic function.

Clock edge triggered logic is far from the ideal of self-timing: all pipeline stages operate in lock-step, with the clock period determined by the slowest path of any pipeline stage. Hence, one could say all but one paths are over-designed. Self-timing, on the contrary, can operate all paths at their maximum speed. Off course, this is a simplified view, as clock edge triggered logic can be optimized for global parameters like area or power more easily, and self-timed logic also needs interfacing between different paths.

When clock edge triggered logic is pushed beyond its margined maximum performance, e.g., higher clock frequency or lower supply voltage, logic paths cannot fully compute. This introduces logic errors. Error tolerant operation has been used in a variety of domains. Error correcting codes are commonly used in telecommunication to overcome channel noise. They rely on adding redundant data in the sender receiver link to detect and possibly correct logic errors introduced along the path [12]. Fault-tolerant operation equally occurs in real-time systems, where hardware components are often duplicated. A majority voter combining the results of these independently operating duplicated components can then significantly reduce the chance of system failure [19]. However, both in telecommunication and real-time systems, hardware overhead and energy consumption are highly inferior to correct operation.

A more circuit-oriented approach are self-checking circuits. Reference [22] presents adder/ALU structure that is capable of self-checking their result by predicting the carry bit of the logic operation. While such techniques require less hardware overhead than full component duplication, their implementation is highly dependent on the logic function to be checked. Additionally, they are hard to implement on control logic. Nonetheless, they have inspired designers to come up with new and innovative circuits providing similar redundant information with less overhead.

5.3.1 Time Redundancy Based Detection

As in logic redundancy checking, time redundancy relies on redundant information to detect and correct errors. In timing redundancy errors are assumed to be soft, meaning there is a deterministic amount of time after which the error perishes. In comparing time-shifted versions of the same signal, errors can be detected. These errors can occur due to single-event upsets like α-particles or V_{dd} droops. The effects of aggressively scaling the supply voltage or the operating frequency are similar: additions in propagation delay due to variations are finite and can thus be tackled in a similar manner with the additional advantage that a later sample is more likely to be correct than an earlier sample.

Reference [23] shows different implementations relying on either a combination of spatial and time redundancy or solely time redundancy. The simplest approach is

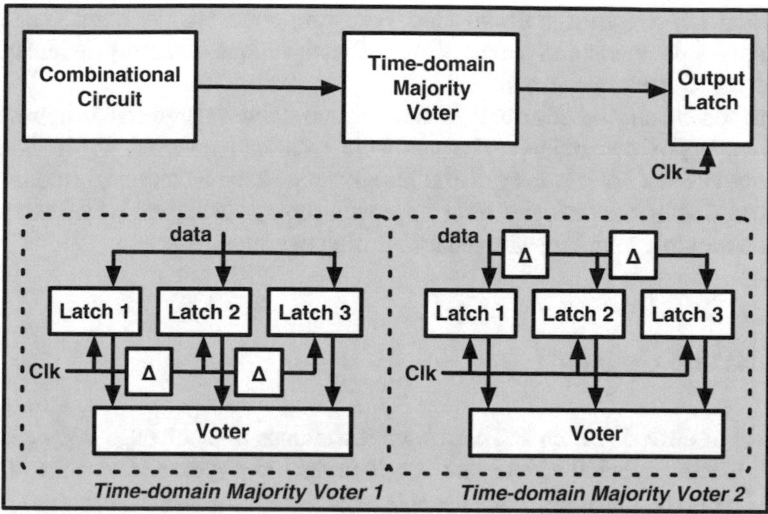

Fig. 5.10 Time-domain majority voter example implementations as shown in [23]

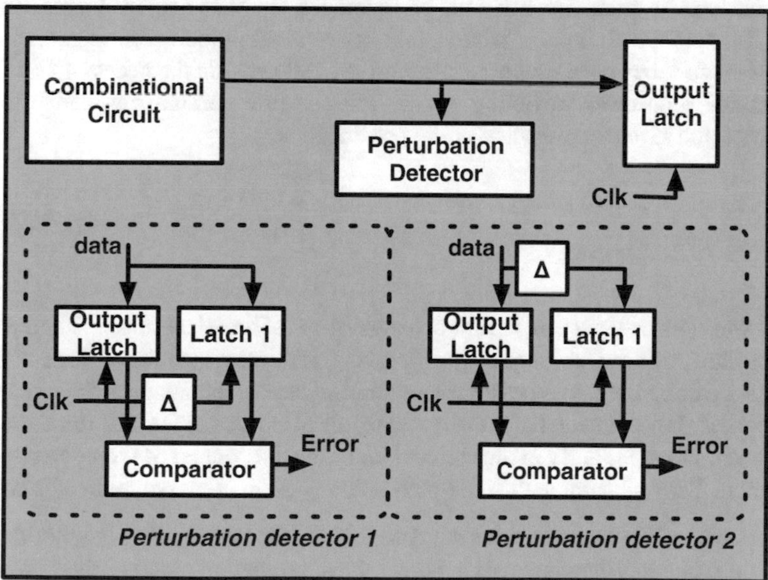

Fig. 5.11 Time-domain perturbation detector implementations as shown in [23]

shown in Fig. 5.10. The result of a combinational circuit is sampled at three different times by either delaying the sampling clock or the combinational result. If the delay is chosen to be at least as long as a possible error perturbation, the majority voter can successfully detect and correct errors. Finally, a more lightweight error detection strategy is shown in Fig. 5.11 in order to detect perturbations. Although it cannot correct errors, it uses significantly less hardware by using the output latch as a reference for comparison. Systems using such a technique often rely on replay/retry procedures to overcome errors.

Time redundancy based error detection is almost always preferred in applications where energy is constrained. The hardware overhead incurred due to multiple sampling is often far less than spatial redundancy based techniques. Additionally, advanced CMOS technologies offer improved timing resolution which allows for accurate multiple sampling with reduced timing overhead.

5.4 EDAC Concepts

The goal of error detection and correction techniques in clock edge triggered logic circuits is to approach the ideal case of self-timing on a system level rather than on a path level. By adjusting either the V_{dd} or the clock frequency, every single die can operate close to, at or beyond its own point-of-first-failure. At this point the performance of each die is maximized, regardless of variations or overall worst case performance. As such, a continuous performance curve is achieved as in Fig. 5.6, much like self-timed circuits, while using conventional clock edge triggered logic. Extensive work has been proposed in literature. All provide the necessary hardware support for robustness to timing errors. This section summarizes their different properties and gives an overview of the possibilities.

5.4.1 Error Strategy

In this work, error detection is first considered in its broadest sense. When relying on real-time resilient techniques to improve worst case operation, three different strategies can be distinguished: error prediction, error detection or error masking. Figure 5.12 shows a schematic overview of all three strategies. All three are real-time in situ strategies and can thus be considered in the context of error detection and correction. They all sample the data-path twice with an identical time shift Δt.

- **Prediction:** The latch samples the data first, while the flip-flop samples the data a time Δt later. When comparing both values, prediction assumes the last sample (flip-flop) to be correct. Thus prediction introduces a margin Δt to prevent errors from being introduced in the data-path: the flip-flop captures and propagates the correct data.
- **Detection:** It uses a similar topology as prediction but the flip-flop samples a time Δt before the latch. As such, the latched value can be considered correct. The data-path flip-flop samples the incorrect data, introducing the need for a data correction strategy. The inherent benefit of this technique is that no margin is introduced: only incorrectly sampled data is flagged as an error.

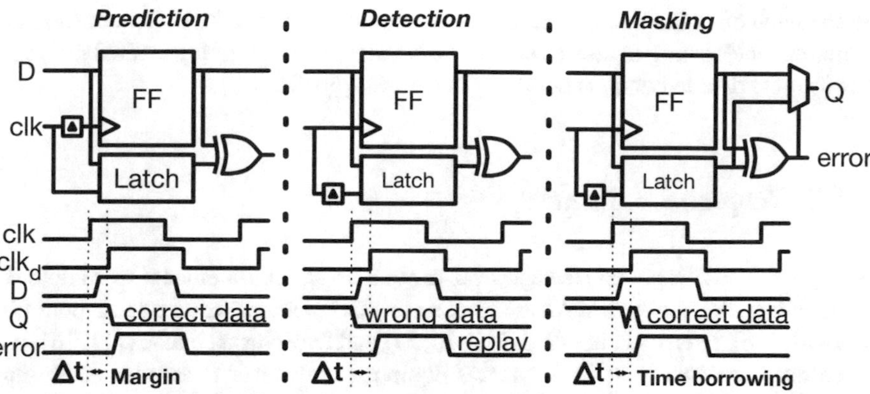

Fig. 5.12 Schematic overview of error prediction, detection and masking with timing diagram implemented with double sampling

- **Masking:** In both prediction and detection, the correct data is readily available because it is used to compare with the sampled value. When making this observation, masking is a straightforward extension. In masking, the error is detected identically as in detection. The error signal can be used to feed the correct data in the data-path. This overcomes the need for a correction strategy. However, it steals a time Δt from the next clock cycle to allow the correct data to arrive and feed through.

The implications at system level of these strategies can be very different. Prediction prevents errors from occurring in any circumstance, thus not requiring error correction. It does however provide real-time in situ information about the near-failure operation of the pipeline. Reference [28] detects transitions occurring in the second half of the clock cycle halfway the data-path. As such, it predicts the onset of an error. Reference [13] prevents errors from occurring by elongating the clock phase when time borrowing events are detected. Error detection detects actual errors, ideally overcoming all margins. It equally provides information about the operation of the pipeline, but relies on a correction mechanism to correct the error. Most systems reported in literature in the last decade employ detection. References [2, 4, 8, 9, 11, 18] are just a few of them. Error masking combines the best of both prevention and detection as it detects actual errors and immediately corrects them. However, it constrains the system in a different way, as will be discussed further in Chap. 6. References [7, 16, 17, 27] are (hybrid) examples of error masking. Reference [7] provides a hybrid approach which allows masking through latch-based time borrowing in predetermined intervals. Reference [16] boosts the supply voltage of the subsequent pipeline stage in the case of a time borrow event, providing the necessary speed-up to overcome a possible timing error. Reference [17] provides the same speed-up by swapping the voltages applied to the device wells, thus employing a simple body biasing scheme. Reference [27] restores the correct data in the latch after the timing window. As such, it maximizes the borrowed time and relies on clock gating to insert a single cycle stall to provide enough calculation time for the new data to compute correctly in the next pipeline stage.

5.4.2 Sequential Element

The previously discussed strategies all rely on sequential elements to sample the data-path at different times. A variety of sequential elements can be used: both flip-flops [4, 7, 8], pulsed latches [2, 3, 7, 9, 26, 27] and two-phased latches [11, 16] have been used to create pipelined systems equipped with error detection. A **flip-flop** based EDAC element relates closest to the conventional flip-flop based pipeline. It captures data on the clock edge and follows much of the same timing constraints as a normal flip-flop based pipeline. It often does increase clock load and area as it adds more sequential and combinational logic. **Pulsed latches** behave similar to a flip-flop: they can be equipped instead of a normal flip-flop when taking in to account their transparency window. This transparency window can be leveraged to enable time borrowing: Reference [9] was the first work to equip pulsed latches exactly for this reason. The pulsed latch allows late data to propagate during the transparency window. What remains is to detect this event. Both EDAC-enabled flip-flops and pulsed latches impose tight short path constraints since they monitor or sample data arriving after the clock edge. Short paths thus require padding to extend their data transitions beyond the monitored window. Depending on the applied detection window, this can have an extensive system area and energy impact. **Two-phase latch** based pipelines as in [11, 16] overcome this requirement as they do not propagate short paths due to non-overlapping clocks. However, conversion of a flip-flop based pipeline to a two-phase latch-based pipeline does introduce significant area and energy overhead. The retiming required in doing this can increase the number of sequential elements significantly, as well as more than double the clock load, resulting in >10% area and energy overhead.

Reference [15] elaborates extensively on the comparison between flip-flop, pulsed latch and two-phase latch in an EDAC multiplier implementation. It shows that sequential area overhead is comparable when using flip-flops or pulsed latches, while two-phased latches increase sequential area significantly. This is mainly due to a customary technique to convert a flip-flop based design into a two-phased latch base design: the master is split from the slave latch and is pushed halfway forward through the data-path. Data-paths tend to have more signalling halfway their data-path, thus requiring more master than slave latches. Additionally, (routing) overhead of having two latches instead of one flip-flop adds to the total overhead. According to Jin et al. [15], combinational area overhead is most impacted by short path padding and is thus influenced by the timing detection window and the short paths adhering to the monitored critical path. Similar detection windows thus result in similar short path padding overhead. No significant differences in combinational area overhead can be expected between flip-flops and pulsed latches.

Pulsed latches benefit from smaller sequential size since they eliminate the master latch of a flip-flop. A clock pulse opens and locks the feedback loop of the latch. Reliable distribution of such a clock pulse can put major constraints on the clock tree, especially in variation-prone conditions. Local pulse generation in the latch [24] or at the lower level nodes of the clock tree [27] may offer a solution. However, the latch propagation delay puts a lower bound on the clock pulse width and slew rate. Since the clock pulse almost always acts as the detection window as well, this prevents timing detection windows of an arbitrary small size and increases the necessary short path padding.

5.4.3 Error Detection Techniques

5.4.3.1 Double Sampling

The easiest way to provide timing redundancy in a clock edge triggered sequential logic based pipeline is by sampling the data at two distinct moments in time: double sampling (DS). It relies on the fact that the logic takes a finite amount of time to compute, making the sample taken at a later time more likely to be correct than the one taken at an earlier time. Typically one sequential element samples first, while another samples Δt later, as shown in Fig. 5.13. When data transitions late, the first element samples the data early, capturing the wrong value. The second element samples later which results in the correct value. Comparison of both values produces an error signal.

Das et al. [8] is one of the first authors implementing such a system in a full processor using a 'Razor flip-flop'. A traditional flip-flop speculatively samples the critical path at the rising edge of the clock. A shadow latch is added to sample the path for a second time, relying on the high clock phase to provide the time shift. Reference [7] applies a similar technique in its TIMBER flip-flop but provides more granularity by having a selective delay for the shadow latch sampling time. The TIMBER latch shown in [7] succeeds in disconnecting master and slave latch of the traditional flip-flop, each sampling the data-path at distinct times and comparing them for error detection. Bubble Razor [11] employs a latch-based pipeline which augments the normal latch with a shadow latch. The shadow latch captures the data-path value when the normal latch becomes transparent, providing a reference for when time borrowing should occur. Kim et al. [17] also used only two latches. It continuously compares the shadow latch value with incoming data, while enabling time borrowing for the main latch. Although double sampling can be considered as a fairly robust detection technique, some problems arise in variation-prone conditions. Especially the mismatch in t_{clk-q} of both sampling elements can introduce additional margin.

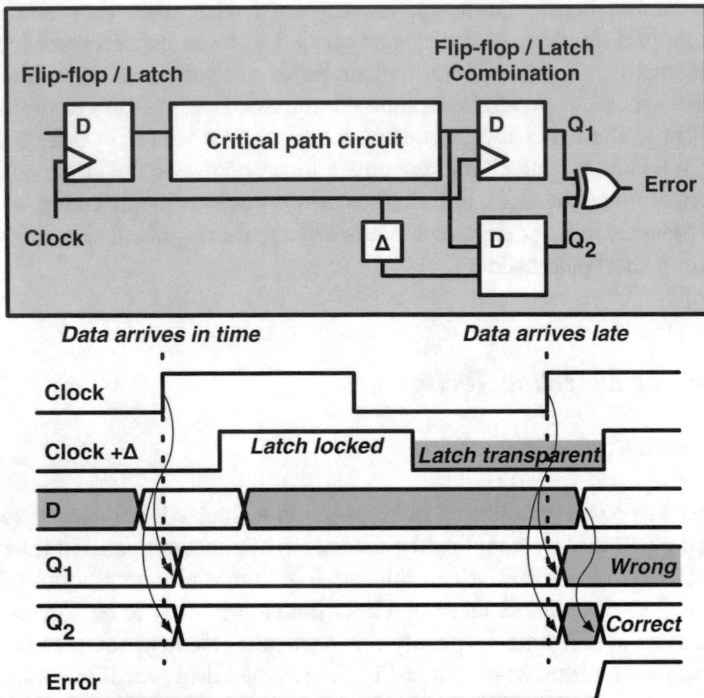

Fig. 5.13 Double sampling principle overview and timing diagram

5.4.3.2 Transition Detection

Another approach to provide timing redundancy is transition detection (TD). A schematic overview is shown in Fig. 5.14. After initially sampling the data-path, any subsequent data-path transitions can be considered as data arriving late, thus errors. In the schematic, late data can still propagate because of the transparent phase of the latch. Less sequential logic is necessary when comparing to double sampling. This relaxes the constraints on the clock network. While also providing redundant data, TD flags any data transition, thus also single-event upsets (SEU) and glitches which otherwise might not be visible using double sampling. Since the TD does not store any information, [2, 9, 13, 26] all use TD augmented with a set-dominant latch (SDL). Note that adding another sequential element again increases overhead. However, the SDL can be shared between multiple latches by bitwise multiplication (OR-operation) of the individual error signals. Zhou et al. [28] detect transitions halfway the data-path on the negative clock edge, thus taking half path delay information as a prediction for full path delay operation.

Virtual supply node monitoring is a convenient way of providing lightweight transition detection, demonstrated in [16, 18, 27]. It can significantly reduce the overhead introduced by transition detection by monitoring the internal nodes of the

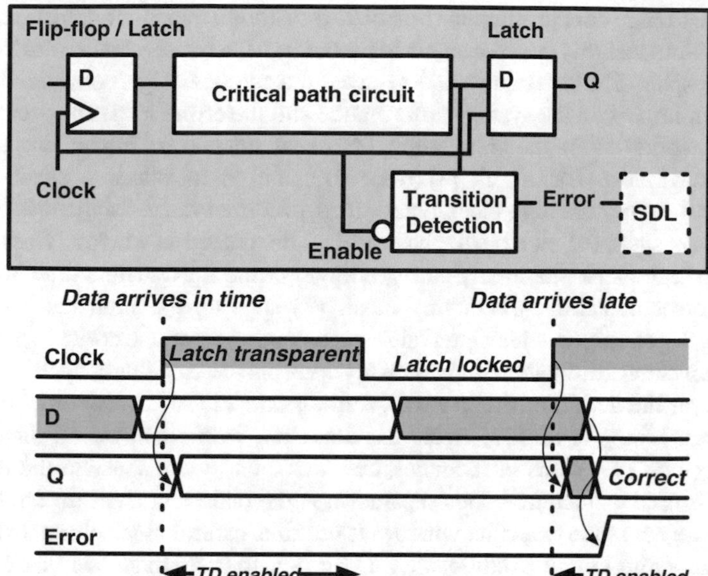

Fig. 5.14 Transition detection principle overview and timing diagram

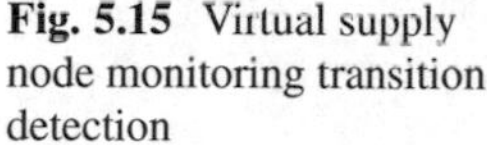
Fig. 5.15 Virtual supply node monitoring transition detection

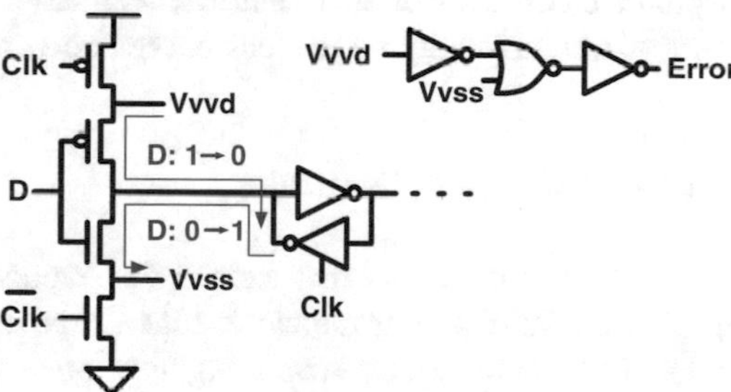

input tri-state inverter. As can be seen in Fig. 5.15 the internal nodes V_{vdd} and V_{vss} (often referred to as side-channels) get charged by the latch feedback loop when data transitions after the clock edge. The internal nodes do experience a V_T drop. Reference [16] extends this technique for ultra-low voltage operation and overcomes the V_T drop by adding complementary devices in parallel.

5.4.3.3 Detection Window

To distinguish between correct and incorrect or in-time and late transitions, all architectures rely on a predefined timing detection window (DW). Hence, the detection window plays a crucial role in the error decision taking. By relying on this detection window, the error detection strategy assumes that the error is transient and has passed within the window time span. As such it can detect the difference in sampled data (double sampling) or data transition (transition detection) in this window. Note that this means every error, whatever the cause, should resolve within this time span. The detection window size is a major design consideration with significant impact on the system performance and the errors it tries to overcome.

Most works equip the transparent phase of latches to implement a timing detection window. Das et al. [8] were the first to introduce a shadow latch, transparent during the high clock phase, next to a traditional rising edge triggered flip-flop. As such, the high clock phase acts as the detection window. Das et al. [9] equip the high clock phase only partly, allowing time for dynamic time borrowing without error detection. Choudhury et al. [7] split up the high clock phase in intervals: depending on which interval flags the timing error, the critical operation of the system can be estimated. Reference [4] controls the clock duty cycle to provide control over the detection window size. References [11, 16, 18, 26] employ similar techniques. Zhang et al. [27] create the detection window based on the clock at multiple (local) locations of the clock tree. As such, it can separate the detection window from the clock tree. This separation is the main argument for not having a clock phase act as the detection window. It requires careful clock phase distribution across the entire circuit. Additionally, it can constrain applications that equip the negative clock edge, such as memory interfaces or other peripherals. Hence, it is generally a good idea to separate the detection window from the clock signal. A separate clock tree for error detection elements can equally achieve this, but often results in significantly more clock energy and area overhead.

5.4.3.4 Hold Time Constraint

Detecting timing errors that exceed the expected clock period inevitably results in data transitions after the clock edge. This results in a conflict with short paths launching data to the same capturing flip-flop as the late arriving data. The concept is illustrated in Fig. 5.16. Two flip-flops feed the same capturing flip-flop through two separate data-paths. The first path is very short and propagates within a fraction of the clock period. The second path is critical and takes the full clock cycle or longer to propagate its data. A timing error flip-flop captures the data. The problem arises in successive clock cycles where both the long path from the previous cycle and the short path from the current cycle manipulate the flip-flop input. Since error detection occurs inside the flip-flop, there is no mechanism to know which path triggered a transition. As such, both transitions will equally be detected and flagged as a timing error. To overcome this, there is no other choice than to pad the short path, increasing its propagation delay to make it extend beyond the detection window. Figure 5.17 demonstrates this. It is evident that this results in an area and energy overhead proportional to the amount of buffers needed to elongate the short path.

While short path padding is the only solution to resolve the hold time problem, its system level overhead can be engineered. It is clear that the amount of short path padding is proportional to the detection window width. Large detection windows consisting of the high clock phase of a 50% duty cycle clock can thus significantly

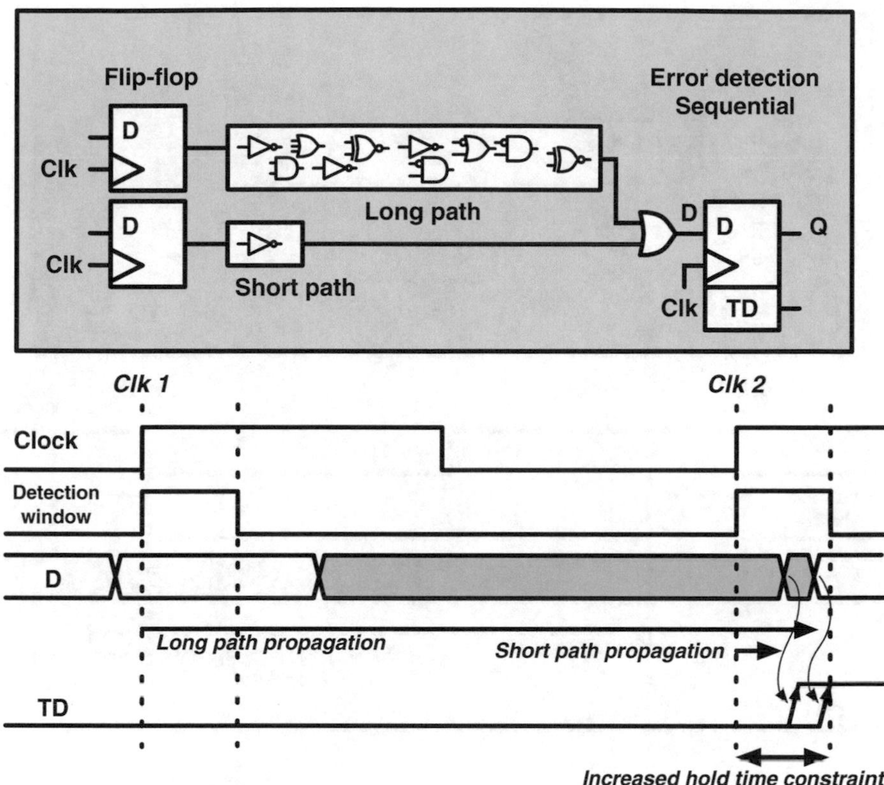

Fig. 5.16 Additional hold time constraint due to error detection after the clock edge

increase the short path padding overhead. Variation sensitivity worsens this effect since it requires margins to guarantee enough padding under variation and PVT conditions. Additionally, this overhead is application dependent. While timing error detectors are typically placed on near-critical paths, the amount of short paths adjoining these near-critical paths is completely unrelated. As a result, there can be a big difference in short path padding overhead between well-controlled pipelined data processing systems and microcontroller-like architectures. Reducing the detection window width is a possible other solution. Kwon et al. [18] tune the detection window post-fabrication to maximize the window with fixed short path padding. References [7, 9] provide an adaptive detection window separately tunable from the clock duty cycle. Reference [21] provides the most optimal detection window by closing the detection window only when data arrives (similar to completion detection). In every case, choosing the detection window width becomes a trade-off between error resilience and short path padding overhead. Both can influence energy consumption. A minimum energy impact solution is preferred, balancing the margins required for error resilience versus those for short path padding.

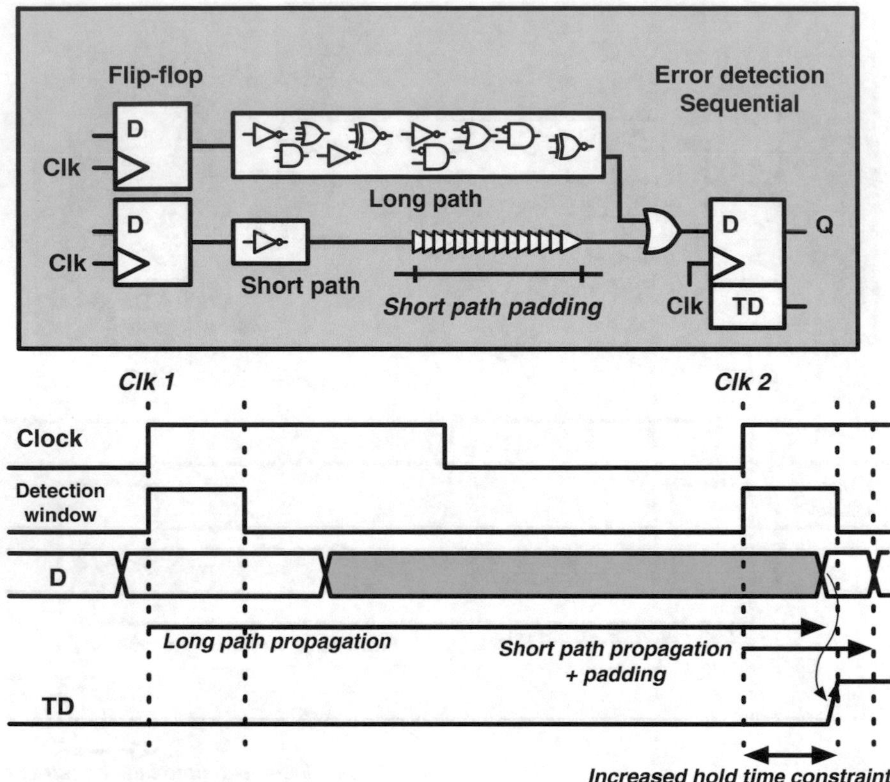

Fig. 5.17 Short path padding to overcome hold time problem due to error detection after the clock edge

5.4.3.5 Metastability

When allowing timing errors to occur in a sequential pipeline, setup timing constraints are inevitably violated. As such, it is possible that the flip-flop becomes metastable. Such event increases the critical path delay as it can linger indefinitely. Additionally, the metastable state can propagate to subsequent stages, resulting in pipeline corruption. Different proposed EDAC strategies can result in different metastability causes. The main considerations are whether metastability can occur at all, and if so, if it can occur in the data-path or the error path.

The research by Das et al. [8] is one of the first works to consider this. Here, metastability can occur in the data-path flip-flop. By attaching a metastability detector to the flip-flop output, it detects such events. The metastability detector consists of a combination of two inverters at the flip-flop output, each with a skewed switching point. The switching points are chosen as to bound the metastable voltage region. The skewness results in two different outputs when their input signal is in the metastable voltage region.

Bowman et al. [2] consider the metastability issue for time borrowing based error detection. Here, the error event is not violating the setup time of the latch since the data can still propagate correctly. Consequently, the metastability problem is removed from the data-path. When using transition detection, the latch that stores the TD event can become metastable, resulting in metastability in the error signal path. This can occur when the pulse generated due to the transition returns to logic low, while the clock signal enabling the transition detector goes to logic high. Metastability in the error path is often considered less problematic since the error signals pass through a well-controlled sequence of logic and flip-flops. These flip-flops act as synchronizers, resolving the metastability issue. Important considerations for metastability are the metastability window (the time frame in which data should arrive to result in metastability) and the metastability resolution time constant. Bowman et al. [2] elaborate on this further, deriving the mean time between potential failures (MTBPF).

Cannizzaro et al. [5] propose a globally asynchronous locally synchronous architecture (GALS) to overcome metastability in EDAC circuits. The key observation here is that the metastable signal always resolves in logic 0 or 1. One of the two is correct value, while the other is erroneous. While a correct value does not trigger error correction, it may result in a significant delay in the data-path because of the time it takes for the metastability to resolve. This can compromise further pipeline stages.

Overall, a good property of time borrowing error detection is the shift in metastability issues to the error path. As this path is often less complex, the risk of metastability reduces significantly. When the time borrowing property is incorporated in the design timing analysis, setup timing constraints can be applied correctly. This prevents metastability. Only when applying dynamic voltage scaling, the setup time can be violated resulting in possible metastability.

5.4.3.6 Sparse Error Detection

Although it is necessary to minimize the overhead introduced by the sequential element, error detection often increases the area and energy of the sequential element significantly. It is not uncommon to result in twice the amount of clock energy and double the flip-flop area. To minimize the overall system impact of such overhead, most EDAC designs take advantage of the timing properties of the system. While a balanced pipeline is the most timing and energy efficient, pipeline imbalance is a reality. This results in timing paths which are less critical than others, hence, less likely to suffer a timing error. A straightforward optimization in reducing the error detection and correction overhead is to only detect timing errors on the paths most likely to report a timing error. Most works implementing an error detection and correction strategy on a processor-like architecture report such sparse insertion of detection circuits. References [4, 8, 9, 16, 18, 26, 27] report replacement ratios between 5% and 30% of the total amount of flip-flops.

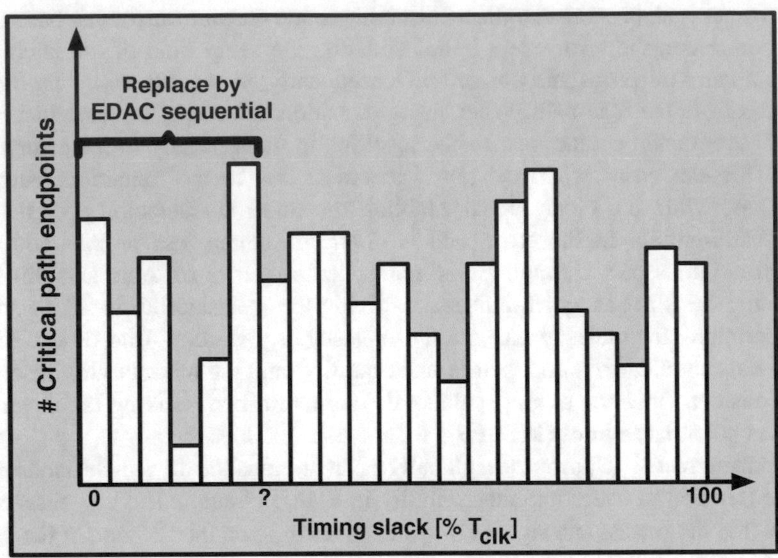

Fig. 5.18 Example timing histogram showing all the path endpoints with their most critical path timing slack

A convenient method for such a consideration is to look at the timing histogram. Figure 5.18 shows such an example histogram. It maps all the sequential endpoints (typically flip-flops) according to the timing slack of the most critical path of which they are the endpoint. In this way, it demonstrates which flip-flops can experience near-critical timing. In sparsely replacing such endpoints with EDAC sequentials, the leftmost endpoints have more chance of experiencing a timing error than their right neighbours. In this consideration, it is wise to consider probabilities rather than pass–fail conditions. Whether a path experiences a timing error depends on its nominal delay but also on the variation, ambient conditions and even the data-sequence it experiences.

In replacing endpoints with EDAC elements, the most important metric is to prevent false positive operation. Under no condition can a timing error occur on a timing path not equipped with error detection, while no error occurs on any error detection enabled path. When considering its probability, the chance of such an event occurring needs to be near-zero. A secondary effect is the resulting energy and area overhead, which is proportional to the amount of endpoints being replaced. A well-balanced pipeline will typically have more near-critical endpoints, hence requires more EDAC-enabled endpoints. It is clear that this effect heavily depends on the architecture and should be evaluated in a case-by-case matter.

5.4.4 Error Correction Techniques

To guarantee functional correctness under all conditions, EDAC strategies employ error correction. While linked to the error detection strategy, correction strategies are often interchangeable. While operating at a lower V_{dd} than worst case conditions allow, it is reasonable to assume that at one point this will result in a timing error. Detecting this timing error has been covered extensively. To guarantee overall system correctness, a correction mechanism is necessary. Correction mechanisms can vary in applicability, area, energy, throughput impact, etc. It is paramount to consider the correction technique carefully when designing an EDAC system. Note that some systems can live with errors by approximating the correct value. Whatmough et al. [26] estimate the correct logic value using a simple algorithm for DSP filter applications.

Majority voting as in [23] and illustrated in Fig. 5.10 can provide error correction by assuming the most occurring value to be correct. This however requires triple redundancy. Tripling the sequential element results in a significant hardware overhead. Recent designs rely on two samples to detect and correct errors. Hence, they consider the last sample to be correct.

The most straightforward technique to correct errors in a processor-like architecture is to use its built-in mechanisms. Instruction replay is such a mechanism. In cached memory systems, instruction replay is used when optimistic instruction scheduling cannot complete due to a cache-miss. Similarly, branch miss-prediction can lead to instructions having to be re-executed. References [2, 4, 9, 18] rely on such mechanism to recover from errors: the current pipeline state is invalidated, while a restore signal triggers re-execution of the faulty instruction. Depending on the cause of the error, simply re-executing the instruction under the same conditions may not warrant correct execution. Aggressive voltage scaling can result in the same error re-occurring. Additional measures can guarantee forward progress: [2, 18] halve the clock frequency, making circuit operation less critical. Bowman et al. [3] also compare this strategy with multiple consecutive executions of the same instruction at nominal clock frequency. Architectural error correction mechanisms are low overhead when they are readily available in the architecture, but come at a high cost when they are not present. Flushing the pipeline comes at a significant throughput cost, and changing the clock frequency can impact synchronous behaviour with adjoining systems. Additionally, these designs cannot recover from timing errors occurring in the pipeline control path. While this is an architecture specific consideration, a generic correction mechanism is preferred.

In doing error detection the correct value is often already present. Das et al. [8] reinsert this value in the pipeline. Figure 5.19 illustrates this principle. The error signal selects the correct data through the MUX in case of an error. This way, the flip-flop samples the correct data the next clock cycle. This mechanism stalls the

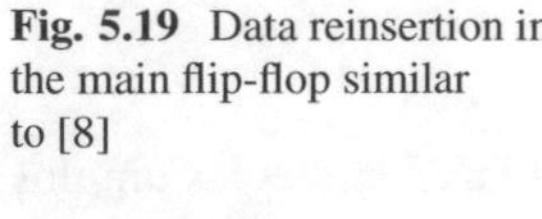

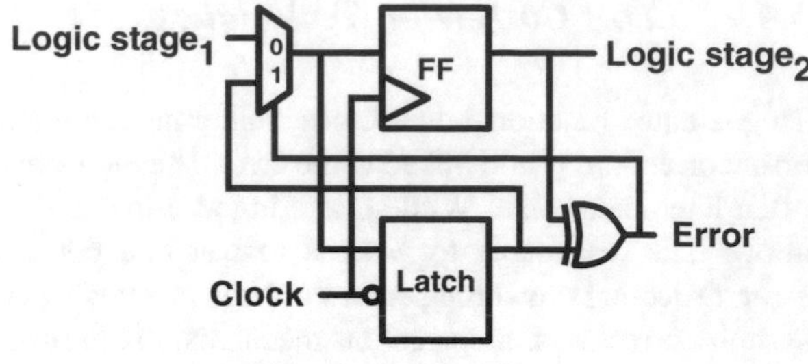

Fig. 5.19 Data reinsertion in the main flip-flop similar to [8]

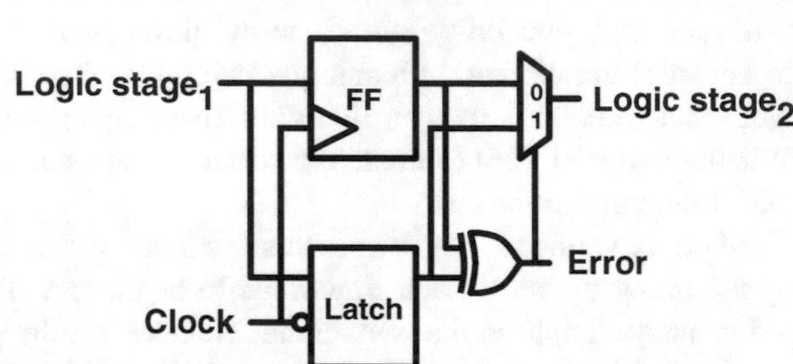

Fig. 5.20 Data reinsertion in the data-path after the main flip-flop

entire pipeline for a single cycle. By consequence, the error signal needs to notify the entire pipeline of this stall before the next clock edge. Note that the late arriving data already consumed part of this clock cycle.

An alternate approach is to reinsert the data after the main flip-flop, as shown in Fig. 5.20. Explicit reinsertion can be realized with a MUX, but implicit reinsertion is more convenient. In this context, time borrowing has to be considered. It is in fact a form of implicitly reinserting the data in the pipeline after the clock edge. A latch allows data to propagate correctly after the clock edge at the cost of an increased hold time constraint and a reduced computation time for the next pipeline stage. What remains is to detect the occurrence of this event, which is often conveniently realized with a transition detector rather than a separate latch. The resulting functionality is very similar to that of Fig. 5.20. The reduced computation time for the next pipeline stage can be a problem. The effective clock period for logic stage 2 in Fig. 5.20 is reduced by the amount of time it lends to logic stage 1 plus the delay of the multiplexer. Choudhury et al. [7] analysed the path delay distribution of an industrial processor and saw few stages became critical despite them lending time to their previous logic stage.

To absolutely guarantee correct computation in subsequent logic stages, most designs that reinsert the data employ a secondary correction mechanism. Reference [11] considers time borrowing events to be errors and stalls instructions in a pipeline counterflow mechanism. Kim and Seok [16] boost the supply voltage of the next logic stage to a higher voltage, resulting in a faster logic propagation: for ultra-low voltage operation, a boost to nominal voltage can result in a 10× speed-up or more. Reference [27] allows for some time borrowing without error detection and only propagates correct data at the end of the detection window. The consecutive clock cycle is gated in order to let the next pipeline stage compute correctly.

Error correction inevitably comes at a cost. Replaying the full pipeline takes in the order of ten clock cycles depending on the pipeline depth. Stalling the clock for a single cycle still compromises throughput as well. Recovering from the error using other techniques can guarantee throughput, but may incur more hardware overhead. Additional error recovery hardware evidently comes at an area and energy cost. For near-threshold operation, the overall trade-off should clearly be expressed in terms of energy. Total energy overhead not only depends on recovery energy, but also the rate by which errors occur. In that context, it is important to study and quantify the timing error causes before implementing the error detection and correction system.

An important consideration to make in correction strategies is its implicit trust in the detection window. While mechanisms to overcome errors are presented, they all presume that the actual timing error can be detected in the detection window time span. This implies that the correct data was available to compare with. Depending on the timing error cause, the necessary timing window to do so can vary.

5.4.5 *PoFF Operation and DVS*

When error resilient operation is achieved, it can be used to reduce energy consumption by operating at a lower V_{dd}. Since voltage scaling reduces dynamic energy consumption quadratically, total system energy often benefits from reduced V_{dd} operation. Figure 5.21 shows the qualitative relationship between supply voltage, energy consumption and error resilience. A margined system requires to be operated

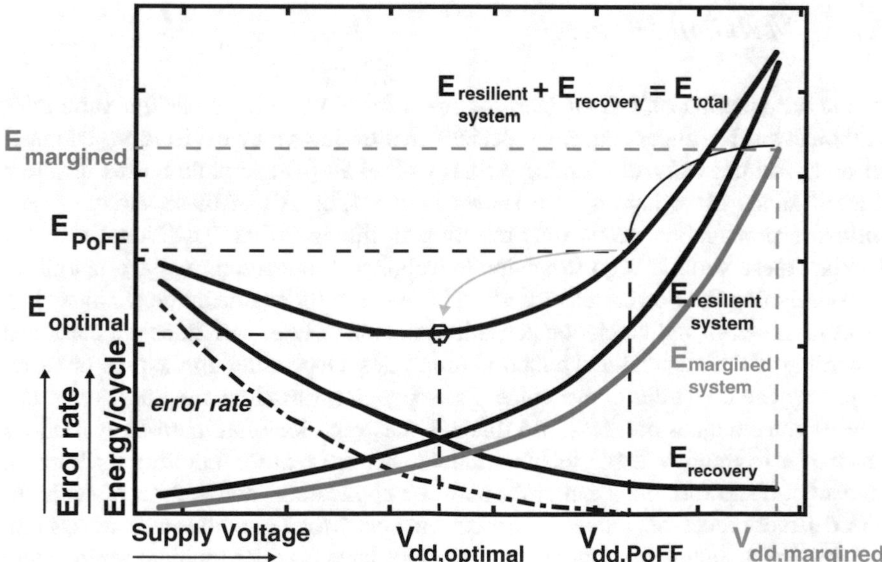

Fig. 5.21 Trade-off in error resilient systems with voltage scaling, similar to [8]

at $V_{dd,margined}$, resulting in an energy consumption denoted as $E_{margined}$. An error resilient system inherently consumes more energy due to the error detection and correction hardware overhead, but allows to decrease the supply voltage. Operation of the resilient system at $V_{dd,PoFF}$ decreases the energy consumption down to E_{PoFF}. In the ideal case E_{PoFF} is lower than the margined operation of the non-resilient system. The key trade-off then is to implement an error detection and correction system at a lower energy cost than margined operation, while also considering the area increase due to such a system. Moreover, it may pay off to operate lower than $V_{dd,PoFF}$ and occasionally correct an error, even at a high energy cost. If errors rarely occur, the energy gain from operation below $V_{dd,PoFF}$ can outweigh the energy cost of an occasional correction procedure. As timing errors increase with lower V_{dd} operation, total error recovery energy increases, adding significantly to the total energy consumption. Systems can rely on this trade-off to reduce overall energy consumption and operate at $V_{dd,optimal}$ and $E_{optimal}$. The trade-off is influenced by almost all parameters available to design an error resilient system: error detection strategy, error correction strategy, error causes, error rate, margined operation, system architecture, etc.

Timing errors can vary in time due to real-time operating conditions, or between dice due to fabrication variations. Because of this, $V_{dd,optimal}$ will vary. Dynamic voltage scaling combined with error information can track this operating point, resulting in near-optimal operation under all conditions.

5.5 ULV Variation-Resilient Operation

5.5.1 Variation Triggers

In the introduction of this chapter, the main focus was on fabrication variations. Although this is a major concern, especially in ultra-low voltage operation, a number of other sources can cause timing errors. Bull et al. [4] elaborate on the different classifications of variations, also shown in Fig. 5.22. All of these variations can influence propagation delay, thus resulting in timing errors. Traditional ways of tackling these variations go from simple techniques like adding voltage margin or over-designing the power delivery network to carefully engineered techniques like process monitors and clock compensation schemes. Here, variations are classified according to their spatial and temporal properties. Depending on the type of variation, they can be harder to overcome. Fast-changing variations require equally fast detection techniques and response times. Local variations like intra-die variations are harder to monitor than global variations, hence in situ monitoring techniques. Impact of these variations can differ between applications and architectures: high-speed server processors induce bigger current spikes than a small sensor subsystem; automotive or industrial applications typically have a wider ambient temperature range then consumer products.

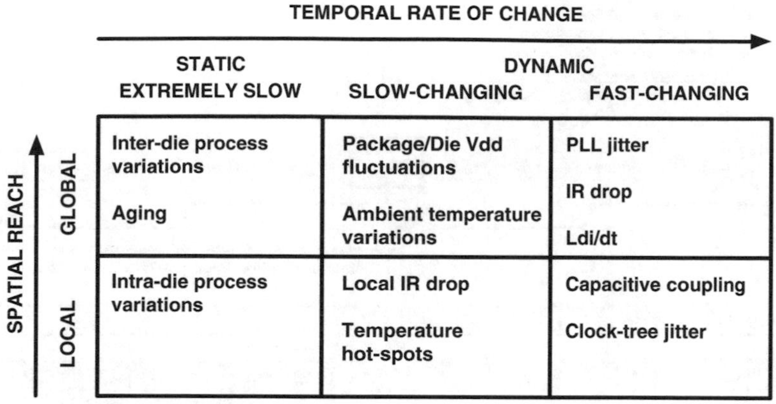

TEMPORAL RATE OF CHANGE →

SPATIAL REACH ↑	STATIC EXTREMELY SLOW	DYNAMIC SLOW-CHANGING	DYNAMIC FAST-CHANGING
GLOBAL	Inter-die process variations Aging	Package/Die Vdd fluctuations Ambient temperature variations	PLL jitter IR drop Ldi/dt
LOCAL	Intra-die process variations	Local IR drop Temperature hot-spots	Capacitive coupling Clock-tree jitter

Fig. 5.22 Classification of variations as in [4]

5.5.2 Variation Mitigation

In situ error detection simply detects timing errors, whatever their cause. Whether it is local or global, slow or fast, any type of variation can result in data-paths not completing within the predetermined clock period. As such, EDAC strategies have been proposed to overcome most of the margins induced by these effects. The in situ monitoring they provide is inherently good for local variations and can thus outperform global monitoring and compensation techniques such as replica biasing. Which margins can be eliminated through error detection and correction depends more on the detection strategy and detection window then on the correction strategy. As most detection strategies rely on the correct data to compare with to flag the erroneous data, the correct data should be available within a predetermined time frame. In this consideration, the detection window is of utmost importance: only timing errors with correct data arriving within the detection window time span can be detected. If a correction strategy is applied, any variation source which meets this condition can be compensated for. As such, a system can be operated reliably with reduced margin. Figure 5.23 illustrates this by distinguishing three cases: normal operation (case 1), a detectable timing error (case 2) and an undetectable timing error (case 3). Any variation source that results in data arrival of path i within the detection window can be overcome using EDAC (case 2). As soon as the variation results in data arrival beyond the detection window ($t_{prop} > T_{clk} + t_{DW}$), the error cannot be detected and reliable operation with reduced margin is not possible, whatever the correction strategy (case 3).

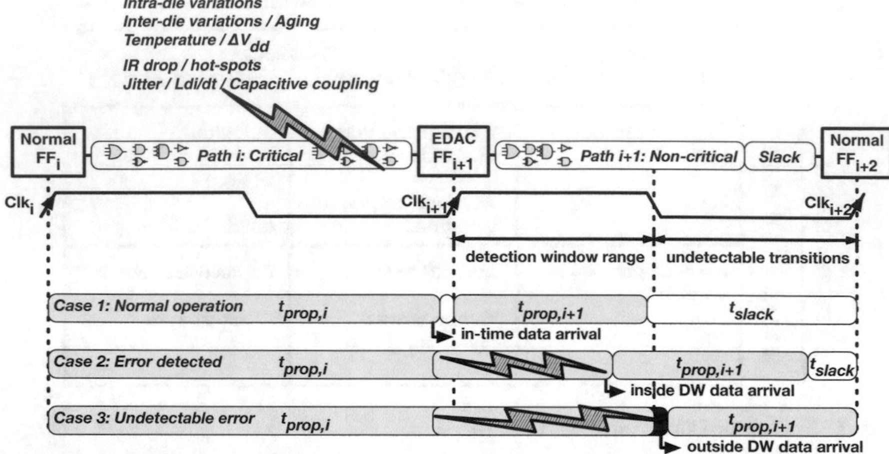

Fig. 5.23 Incident variation resulting in a possible timing error. Case 1: normal operation. Case 2: incident variation results in detectable error. Case 3: incident variation results in an undetectable error

5.5.3 Ultra-Low Voltage

Performance of ultra-low voltage operated systems exhibits more variation induced shifting. Enabling error detection and correction can thus improve these systems significantly, benefiting optimally from the intended energy reduction. While conservative voltage margins result in a huge overhead for ultra-low voltage systems, error resilient operation equally results in overhead. Especially since the error detection and correction circuits suffer from the same variation, hence also requiring some margin for robust operation.

In that sense, it is a real challenge to design error resilient systems. Not only do the circuits have to function at ultra-low voltage, they also need to be robust enough to improve on traditional techniques. Error resilient designs often overcome most of the margins on the data-path. What is less clear is the margins they add in the error detection and correction path. For example, the double sampling technique illustrated in Fig. 5.13 relies on instantaneous sampling at two distinct moments in time. At ultra-low voltage, the sample delay Δt may incur such uncertainty that the time of sampling varies significantly. Uncertainty in t_{clk-q} of both sequential elements also adds to the unpredictability of the data-path sample. In these cases, there is no other option then to counteract this uncertainty by careful design and enough margin to guarantee correct operation. From a system point of view, a false negative (error is detected while there was none) is much more tolerable than a false positive (no error was detected while there was one). Every false negative can be considered as margined overhead, as energy is consumed due to correction while there was no actual error. The margins necessary for robust ultra-low voltage operation are most prominent in the detection window, the short path padding and the error correction path.

5.5.3.1 Detection Window

Robust generation of the detection window (DW) is similar to the robust generation of a clock. Variation-resilient distribution of this DW to a leaf node is a similar challenge as clock tree distribution. Any jitter in the DW generation or signal distribution results in uncertainty on the error detection. Careful DW generation requires large devices since they are less intra-die variation prone, as does DW signal distribution. Figure 5.24 shows a schematic overview of how root node detection window separated from the clock generation is influenced by variation. The clock leaf node and DW node vary independently due to uncorrelated influence of device fabrication variation and parasitic effects. This results in a large false negative error detection region and a small effective detection window: a large part of the detection window precedes the clock edge, resulting in overly conservative error detection.

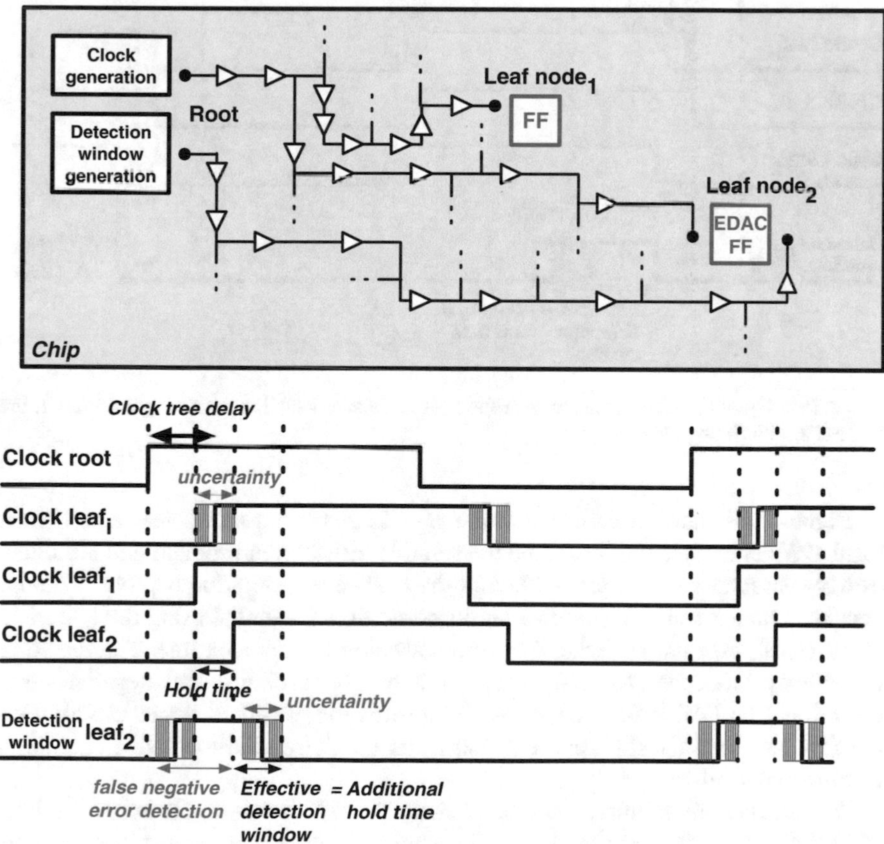

Fig. 5.24 Diagram of clock and detection window generation at the root node with timing diagram

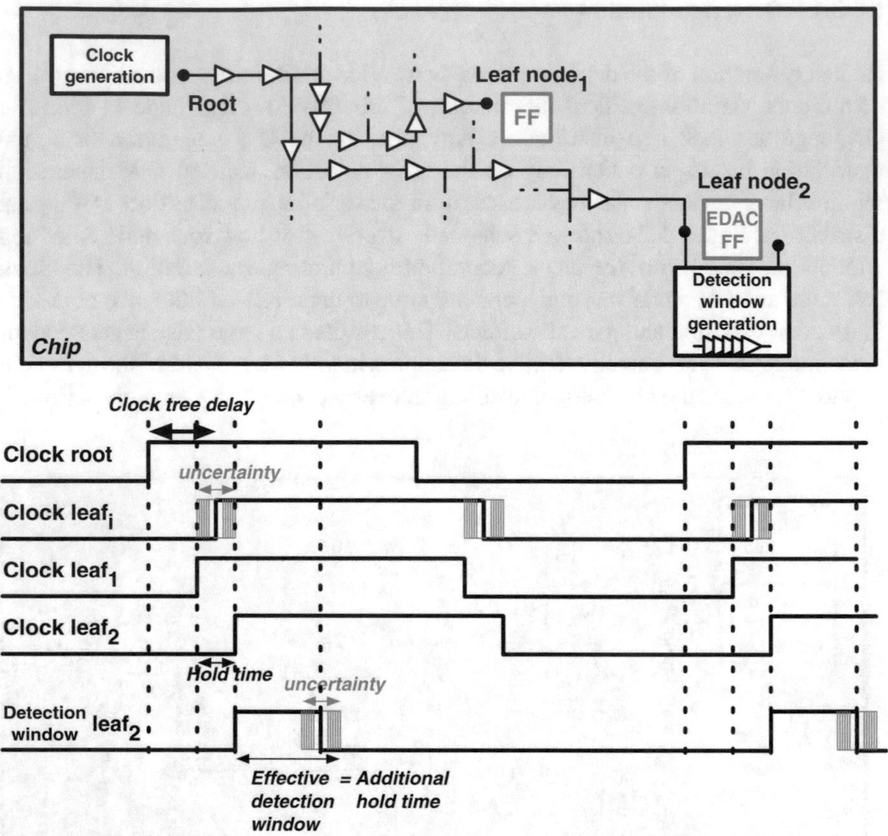

Fig. 5.25 Diagram of clock generation at the root node and detection window generation at the leaf node, with timing diagram

Figure 5.25 shows a case where the DW is generated at the leaf node. Such local DW generation can overcome the reliable distribution problem and eliminate overly conservative detection. Although the leaf node generation is probably more area constrained and thus more variation prone, it is generated using the leaf node clock signal, thus not suffering from any variation in the clock tree. The resulting uncertainty influences the DW size, but does not result in false negative error detection. The DW is fully effective. While this may result in some variation on the DW size between different leaf nodes, its relation to the local clock signal is carefully controlled.

An intermediate solution could be to share the DW generation between a subset of leaf nodes. In that case the leaf nodes should be chosen to be as close together as possible and share as much of the clock tree as possible. More generally, a larger common path between the DW and the clock signal results in less uncertainty, similar to the conclusion presented in Sect. 4.3.3 on clock tree synthesis.

5.5.3.2 Short Paths

Short path padding may require significant margins at ultra-low voltage. Positive hold time is mainly a consequence of clock arrival mismatch between the launching and capturing flip-flop. In this regard, it benefits from reliable distribution of the clock signal. The effect is shown in the timing diagrams of Figs. 5.24 and 5.25. In EDAC circuits, the DW adds to the hold time constraint. Thus, uncertainty on the detection window size results in uncertainty of the hold time constraint, hence margins. As hold time is a relationship between two separate leaf nodes, worst case conditions should be assumed. Local DW generation results in a larger effective detection window, hence a larger hold time. Local DW generation is also variation prone due to its constrained area. This increases the DW uncertainty and thus increases the amount of necessary short path padding. However, short path padding also depends significantly on the architecture and how the short paths relate to the critical paths. To facilitate a generic and well-controlled ULV approach, local detection window generation is still preferred because of its larger effective detection window.

5.5.3.3 Correction Path

The error correction path is the third aspect that may require significant margining. Although the exact approach may vary between different error correction strategies, error signals are often used to control architecture-level signals. Previously discussed examples include gating the clock signal at the clock tree root, the clock leaf nodes or stalling the pipeline in a counterflow manner. These techniques all rely on robust and fast distribution of such signals to the clock root or to other pipeline stage leaf nodes. Since error detection inherently occurs after the clock edge, a sequential system can only equip the remaining time of the clock period to propagate the error signals.

The principle is illustrated in Fig. 5.26. Generating the global *error*-signal from the local *error*-signals requires a bitwise OR-operation of all the error signals, resulting in a huge fan-in. Some implementations rely on a faster dynamic OR-gate for this operation. It is clear such logic topology is unsuitable for ULV operation. Additionally, the timing variation on this long logic path may result in timing-critical operation. This is unacceptable: while the goal was to flag timing-critical operation of one path, in this case the error signal may be lost despite being correctly detected, just by propagating it up the architecture. Again, this problem depends heavily on the architecture. The amount of *error*-signals to be consolidated and the propagation time through the architecture depend on the amount of near-critical paths and the architecture size.

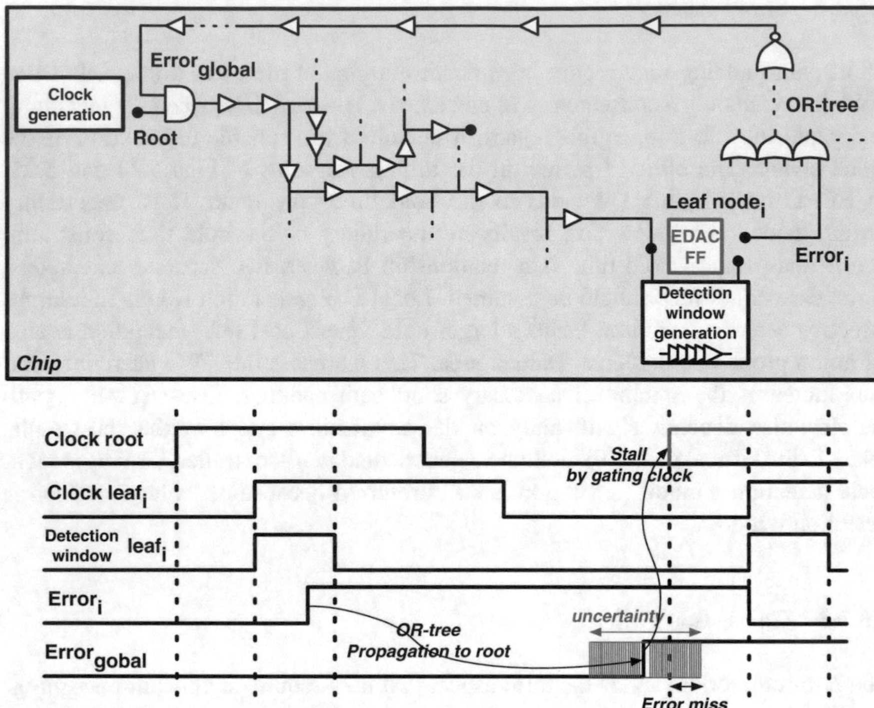

Fig. 5.26 Diagram of error signal propagation to the clock root node with timing diagram

5.6 Baseline Comparison

It is clear from Sect. 5.4.5 that the goal of implementing an EDAC strategy is improving the overall design either in energy consumption or in operating frequency. This chapter has shown that this is non-trivial, as a variety of factors are influenced by the EDAC design choices. However, on what do they improve? A good baseline comparison is necessary. One could argue that the ideal EDAC design has no baseline, as a full vertical integration of all the considerations definitely leads to the best results. However, with the recent increase in EDAC-like publications, evaluating the quality of a strategy is necessary. Comparison with other state-of-the-art work definitely makes sense from a qualitative perspective. The answer to the previous considerations heavily depends on the application, the architecture and the source of the variations to be tackled. Comparing absolute numbers often makes less sense. To demonstrate the improvement of their EDAC strategy, authors usually provide a baseline (measurement) to compare with.

5.6.1 Ideal Baseline Design

In any context, it is best to consider a separate fully optimized design without EDAC as the baseline implementation. Some EDAC strategies use pulsed latches or two-phased latches to facilitate error detection and correction. As most industrial designs still equip a conventional flip-flop based pipeline, the baseline implementation should also. Especially when a non-flip-flop based design would suffer some overhead, as it does with pulsed latches or two-phased latches. The baseline design should be signed off according to industry standards, thus equipping margin for different sources of variations. As the sign-off point provides the speed performance, power simulations in different corners at this speed should be reported. Ideally, silicon measurements across multiple dice coming from different corners under a variety of ambient conditions paint the clearest picture of the baseline performance.

5.6.2 Other Baseline Comparisons

The ideal baseline pictured in the previous section can be hard to report. A separate silicon implementation can come at a high cost. For this reason, authors often provide other additional comparisons. The most reported comparison cases are a margined and an unmargined baseline measurement. Some authors report simulation results of these cases instead.

A margined baseline design can be subjective. The minimum amount of margin to overcome certain sources of variation can be hard to determine. For this reason, some authors (as does Chap. 6 of this work) report the unmargined baseline performance. Such measurements provide the *ideal* performance in both speed and energy consumption, overcoming all margins at no cost. These measurements can be achieved through extensive silicon testing to reach f_{max} and E_{min} performance across a wide V_{dd} range. In doing so, it allows to report the net overhead due to the EDAC implementation.

Some EDAC implementations allow operation without error detection. Hence, some authors report measurements of the EDAC-enabled silicon, but with the error detection and correction disabled. Margined or unmargined measurements of such cases are found in literature. Depending on how the EDAC operation can be disabled, speed performance of such a test case can be similar or not to a baseline design. Energy consumption is definitely overestimated, since overhead due to short path padding (both static and dynamic energy) or error detection element (mostly static energy) cannot be eliminated during these measurements.

While the ideal baseline is always preferred, it is clear that such comparison is not always possible. Figure 5.27 helps to navigate the different baseline comparisons and their energy breakdown and improvement. An encouragement to authors would be to clearly report the baseline conditions, while readers should always carefully consider the reported baseline when concluding on the reported work.

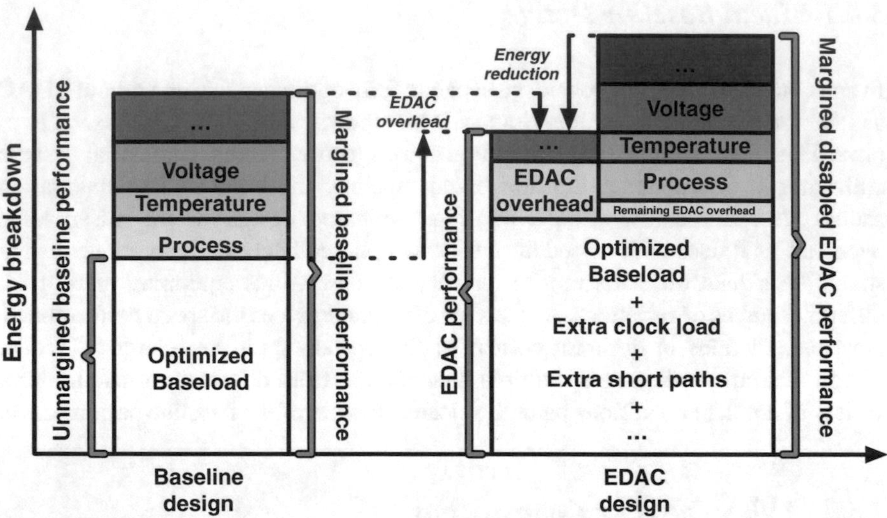

Fig. 5.27 Energy breakdown of baseline comparison EDAC implementations

5.7 Conclusion

This chapter gave an overview of the most important considerations in error detection and correction circuits. Some of it is summarized in Fig. 5.28. Since many error detection and correction strategies have been reported in literature, it is hard to navigate their improvements and drawbacks. First and foremost the kind of variations to target and their influence should be clear. What are traditional ways of tackling these and can error detection and correction overcome them. Next, a robust error detection element should be available. It should operate at the targeted voltage range and should integrate in the design flow. This often closely relates to the error detection technique, its influence on short path padding, metastability and clock/detection window generation. A correction strategy in line with the variations to be tackled and cooperating with the error detection circuits is necessary. Finally, architectural considerations are equally important. Some systems allow easier error detection and correction integration than others. The entirety of these considerations maps into the overall comparison: does the EDAC-enabled system improve on the original margined system in operating frequency, energy consumption or both.

For ultra-low voltage implementation, EDAC overhead can be in line with the margins necessary to enable predictable operation. This means it can come at a non-negligible cost. Both transition detection and double sampling have pro's and con's when considering ULV operation. Transition detection requires less sequential elements for detection and performance does not degrade due to mismatch between both sampling elements. They both rely on a detection window which should be robust enough to enable correct EDAC operation. Short path padding equally benefits from a carefully controlled detection window. Local detection window

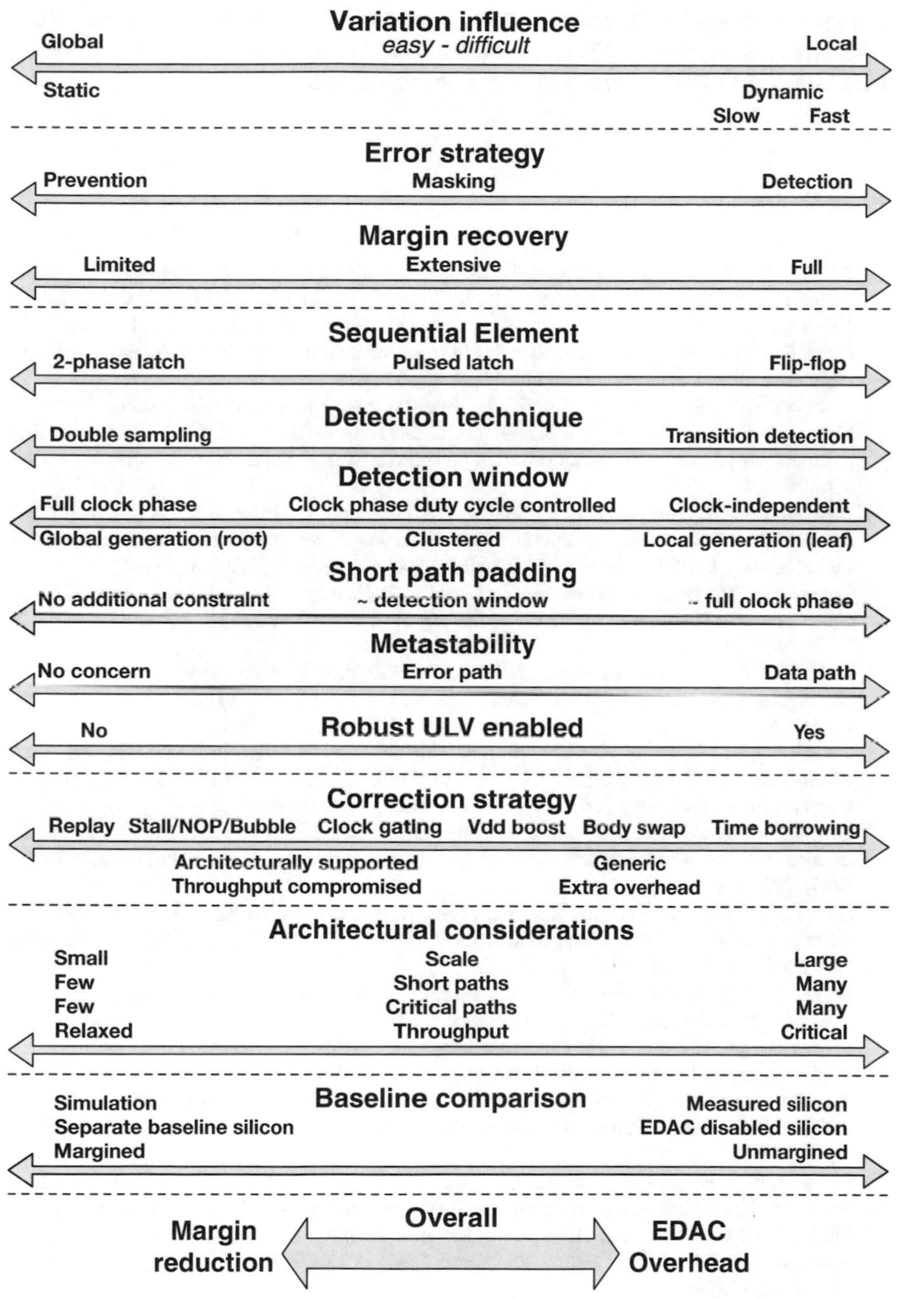

Fig. 5.28 Summary of the EDAC properties, considerations and challenges

generation provides a good middle ground between detection window effectiveness and hold time uncertainty. To minimize the error correction overhead, error masking is a good strategy. Inherent correction also makes the system level error correction signal less timing critical. These considerations are put to the test in Chap. 6, where a ULV-enabled EDAC microcontroller is implemented.

References

1. Akgun, O.C., Leblebici, Y., Vittoz, E.A.: Design of completion detection circuits for self-timed systems operating in subthreshold regime. In: PRIME, pp. 241–244. IEEE, Piscataway (2007)
2. Bowman, K.A., Tschanz, J.W., Kim, N.S., Lee, J.C., Wilkerson, C.B., Lu, S.L.L., Karnik, T., De, V.K.: Energy-efficient and metastability-immune resilient circuits for dynamic variation tolerance. IEEE J. Solid-State Circuits **44**(1), 49–63 (2009)
3. Bowman, K.A., Tschanz, J.W., Lu, S.L.L., Aseron, P.A., Khellah, M.M., Raychowdhury, A., Geuskens, B.M., Tokunaga, C., Wilkerson, C.B., Karnik, T., De, V.K.: A 45 nm resilient microprocessor core for dynamic variation tolerance. IEEE J. Solid-State Circuits **46**(1), 194–208 (2011)
4. Bull, D., Das, S., Shivashankar, K., Dasika, G.S., Flautner, K., Blaauw, D.: A power-efficient 32 bit ARM processor using timing-error detection and correction for transient-error tolerance and adaptation to PVT variation. IEEE J. Solid-State Circuits **46**(1), 18–31 (2011)
5. Cannizzaro, M., Beer, S., Cortadella, J., Ginosar, R., Lavagno, L.: SafeRazor: metastability-robust adaptive clocking in resilient circuits. IEEE Trans. Circuits Syst. I Regul. Pap. **62**(9), 2238–2247 (2015)
6. Cheng, F.C.: Practical design and performance evaluation of completion detection circuits. In: Proceedings International Conference on Computer Design. VLSI in Computers and Processors, pp. 354–359. IEEE Computer Society, Washington (1998)
7. Choudhury, M., Chandra, V., Mohanram, K., Aitken, R.: TIMBER: time borrowing and error relaying for online timing error resilience. In: IEEE Design, Automation & Test in Europe Conference & Exhibition (DATE), pp. 1554–1559. EDAA, Dresden (2010)
8. Das, S., Roberts, D., Lee, S., Pant, S., Blaauw, D., Austin, T., Flautner, K., Mudge, T.: A self-tuning DVS processor using delay-error detection and correction. IEEE J. Solid-State Circuits **41**(4), 792–804 (2006)
9. Das, S., Tokunaga, C., Pant, S., Ma, W.H., Kalaiselvan, S., Lai, K., Bull, D.M., Blaauw, D.T.: RazorII: in situ error detection and correction for PVT and SER tolerance. IEEE J. Solid-State Circuits **44**(1), 32–48 (2009)
10. Dean, M., Dill, D., Horowitz, M.: Self-timed logic using current-sensing completion detection (CSCD). In: IEEE International Conference on Computer Design: VLSI in Computers and Processors, pp. 187–191. IEEE Computer Society Press, Washington (1991)
11. Fojtik, M., Fick, D., Kim, Y., Pinckney, N., Harris, D.M., Blaauw, D., Sylvester, D.: Bubble Razor: eliminating timing margins in an ARM cortex-M3 processor in 45 nm CMOS using architecturally independent error detection and correction. IEEE J. Solid-State Circuits **48**(1), 66–81 (2013)
12. Gallager, R.: Low-density parity-check codes. IEEE Trans. Inf. Theory **8**(1), 21–28 (1962)
13. Hiienkari, M., Teittinen, J., Koskinen, L., Turnquist, M., Kaltiokallio, M.: A 3.15pJ/cyc 32-bit RISC CPU with timing-error prevention and adaptive clocking in 28nm CMOS. In: Proceedings of the IEEE Custom Integrated Circuits Conference (CICC), pp. 1–4. IEEE, Piscataway (2014)
14. Hing-mo, L., Tsui, C.Y.: High performance and low power completion detection circuit. In: Proceedings of the International Symposium on Circuits and Systems (ISCAS), vol. 5, pp. 405–408. IEEE, Piscataway (2003)

15. Jin, W., Kim, S., He, W., Mao, Z., Seok, M.: In situ error detection techniques in ultralow voltage pipelines: analysis and optimizations. IEEE Trans. Very Large Scale Integr. (VLSI) Syst. **25**(3), 1032–1043 (2017)
16. Kim, S., Seok, M.: Variation-tolerant, ultra-low-voltage microprocessor with a low-overhead, within-a-cycle in-situ timing-error detection and correction technique. IEEE J. Solid-State Circuits **50**(6), 1478–1490 (2015)
17. Kim, S., Cerqueira, J.P., Seok, M.: A 450mV timing-margin-free waveform sorter based on body swapping error correction. In: IEEE Symposium on VLSI Circuits (VLSI-Circuits), pp. 1–2. IEEE, Piscataway (2016)
18. Kwon, I., Kim, S., Fick, D., Kim, M., Chen, Y.P., Sylvester, D.: Razor-lite: a light-weight register for error detection by observing virtual supply rails. IEEE J. Solid-State Circuits **49**(9), 2054–2066 (2014)
19. Lyons, R.E., Vanderkulk, W.: The use of triple-modular redundancy to improve computer reliability. IBM J. Res. Dev. **6**(2), 200–209 (1962)
20. Mead, C., Conway, L.: Introduction to VLSI Systems. Addison-Wesley Publishing, Reading (1980)
21. Nejat, M., Alizadeh, B., Afzali-Kusha, A.: Dynamic flip-flop conversion: a time-borrowing method for performance improvement of low-power digital circuits prone to variations. IEEE Trans. Very Large Scale Integr. (VLSI) Syst. **23**(11), 2724–2727 (2015)
22. Nicolaidis, M.: Efficient implementations of self-checking adders and ALUs. In: 23rd International Symposium on Fault-Tolerant Computing, pp. 586–595. IEEE Computer Society Press, Washington (1993)
23. Nicolaidis, M.: Time redundancy based soft-error tolerance to rescue nanometer technologies. In: Proceedings 17th IEEE VLSI Test Symposium, pp. 86–94. IEEE Computer Society Press, Washington (1999)
24. Partovi, H., Burd, R., Salim, U., Weber, F., DiGregorio, L., Draper, D.: Flow-through latch and edge-triggered flip-flop hybrid elements. In: IEEE International Solid-State Circuits Conference Digest of Technical Papers (ISSCC), pp. 138–139. IEEE, San Francisco (1996)
25. Uht, A.: Going beyond worst-case specs with TEAtime. IEEE Micro Top Picks **37**(3), 51–56 (2004)
26. Whatmough, P.N., Das, S., Bull, D.M.: A low-power 1-GHz razor FIR accelerator with time-borrow tracking pipeline and approximate error correction in 65-nm CMOS. IEEE J. Solid-State Circuits **49**(1), 84–94 (2014)
27. Zhang, Y., Khayatzadeh, M., Yang, K., Saligane, M., Pinckney, N., Alioto, M., Blaauw, D., Sylvester, D.: iRazor: current-based error detection and correction scheme for PVT variation in 40-nm ARM Cortex-R4 processor. IEEE J. Solid-State Circuits **53**(2), 619–631 (2018)
28. Zhou, J., Liu, X., Lam, Y.H., Wang, C., Chang, K.H., Lan, J., Je, M.: HEPP: a new in-situ timing-error prediction and prevention technique for variation-tolerant ultra-low-voltage designs. In: IEEE Asian Solid-State Circuits Conference (A-SSCC), pp. 129–132. IEEE, Piscataway (2013)

Chapter 6
Timing Error-Aware Microcontroller

Abstract The error detection and correction design space was presented in Chap. 5. It discussed the different implementation considerations to be made when realizing a state-of-the-art EDAC system. As such, it created a framework to navigate when analysing or implementing an effective EDAC system. A key observation when looking at state-of-the-art is that few EDAC strategies enable ultra-low voltage operation. Considering that ultra-low voltage designs are most susceptible to over-design, EDAC operation can be a key enabler for effective ultra-low voltage systems.

This chapter does exactly that: it implements a state-of-the-art ultra-low voltage error detection and correction enabled microcontroller. It uses in-flip-flop transition detection controlled by a locally generated detection window. At the same time, the master and slave latch of the flip-flop are clocked independently to realize a transparency window that allows error masking similar to a latch. This way, data arriving after the clock can still propagate correctly while being flagged as timing errors. A system level error processor helps to control the autonomous dynamic voltage scaling loop that realizes point-of-first-failure operation.

In Sect. 6.1, the different EDAC considerations for ULV are discussed and the proposed EDAC strategy is presented. Section 6.2 presents the circuits that realize ULV error detection and correction and discusses their near-threshold operation. Section 6.3 demonstrates how these circuits are modelled and how the EDAC strategy is integrated in a generic VLSI design flow. The developed prototype is extensively presented in Sect. 6.4: the EDAC strategy is applied to a microcontroller architecture similar to the ones presented in Chap. 4. It demonstrates how the architecture influenced the design decisions and what was necessary to implement an effective system. Section 6.5 presents the measurement results of the silicon implementation and how they compare to the baseline system presented in Chap. 4. Section 6.6 compares the proposed system to other strategies and state-of-the-art implementations. Finally, Sect. 6.7 draws a conclusion on the proposed EDAC strategy and prototype.

H. Reyserhove, W. Dehaene, *Efficient Design of Variation-Resilient Ultra-Low Energy Digital Processors*, https://doi.org/10.1007/978-3-030-12485-4_6

6.1 Architecture

This work proposes timing error detection and masking through time borrowing. The goal is to detect late data transitions[1] caused by reduced guard band operation. As they would normally result in faulty operation, such critical operation is impossible without error detection and correction infrastructure. If timing error events are detected and overcome, resilient operation is enabled. Additionally, error information can guide dynamic voltage scaling to optimize reduced guard band operation. Such closed loop operation can overcome much of the energy overhead associated with ULV operation.

The EDAC flip-flop consists of five key elements (see Fig. 6.1): a soft-edge flip-flop, a transition detector, a timing/control block, an error latch and an error processor. The combination of these five building blocks can be applied to any traditional flip-flop based design to enable error detection and correction, thus improving its energy efficiency. The operation is as follows: the transition detector detects timing error events in the flip-flop and stores these in the error latch. The soft-edge flip-flop enables error detection and inherently corrects these errors as it allows data arrival after the clock edge. It thus not requires architecture level error correction. The timing/control block generates the necessary signals derived from the clock to facilitate this operation. Finally, the error processor consolidates the error flags coming from the error latches across the entire design and processes

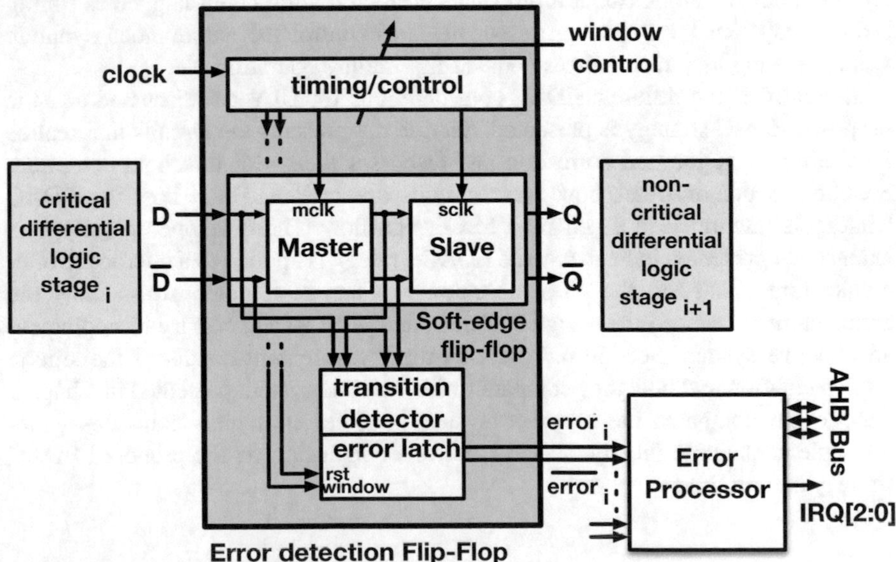

Fig. 6.1 Diagram of the proposed EDAC architecture

[1] Further referred to as timing errors.

them. Subsequently it uses the error information to decide on dynamic voltage scaling and better-than-worst-case operation.

6.1.1 Transition Detection

Timing errors are detected using a robust transition detector. Most transition detectors rely on comparing two time-shifted signals of the flip-flop input data: a transition results in a logic level difference of the two. To create a time-shifted version of the data input, some architectures rely on additional logic (see Sect. 5.4.3.2). This results in a hardware overhead, since the additional logic is only required to create the necessary delay. As shown in Fig. 6.1 this work equips the master latch of the flip-flop as a delay element to enable transition detection. As the master latch is transparent, incoming data transitions experience the delay of the master latch. The transition detector in this work leverages this delay to flag that transition.

In case of a timing error, the data transition occurs after the clock edge. In a normal master/slave flip-flop, the master latch would be locked after sampling erroneous data. Using a normal master/slave flip-flop thus prevents such transition detection. This work uses a soft-edge flip-flop, thereby enabling post-clock edge transition detection over the master latch.

6.1.2 Soft-Edge Flip-Flop

A soft-edge flip-flop [5, 9] differs from a normal flip-flop in its most basic property: the edge-triggered sampling. As the name says, the soft-edge flip-flop does not sample at the clock edge. Rather, it samples the data a finite amount of time later. In the time period bounded by the clock edge and the actual sampling time, the flip-flop is transparent. As such, it behaves exactly like a latch: it propagates data during its transparent phase and samples (locks) the data at the end of that phase. The way the soft-edge flip-flop achieves this operation differs from a latch: it still equips a master and slave latch, but controls them with a time-shifted clock. This results in some overlap in the transparent phase of both latches, enabling the operation as described. Figure 6.1 shows that the soft-edge flip-flop is controlled using a separate master ($mclk$) and slave clock ($sclk$).

An important consideration for ultra-low voltage operation is that the soft-edge flip-flop, as opposed to a normal latch, can enable a very small transparent phase. Controlling a latch with a duty cycle modulated clock (thus a pulsed latch) can theoretically achieve the same operation. However, locking the data in the latch requires a minimum pulse period (minimum pulse width constraint). Additionally, distributing a small pulse width clock signal robustly in the system can prove to be difficult. In a soft-edge flip-flop the transparency window is limited only by the delay element necessary to create a reliable delay between the master and slave clock. It can thus operate with an arbitrary small transparency window.

6.1.3 Timing/Control

To create a transparency window in the soft-edge flip-flop and control the sampling property of the error latch, the timing/control block creates some signals derived from the *clock* signal. The slave latch of the soft-edge flip-flop is controlled using the original clock, while the master latch receives a delayed clock. The timing/control block delays the clock using a controllable delay line. The overlap of both clocks constitutes the transparency window of the flip-flop and determines when the transition detection should be active. The timing/control block therefore enables the error latch using the *window* signal, which is derived from the original and delayed clock. Figure 6.2 shows the logic operations on the *clock* signal resulting in the different clocks and control signals necessary for EDAC operation.

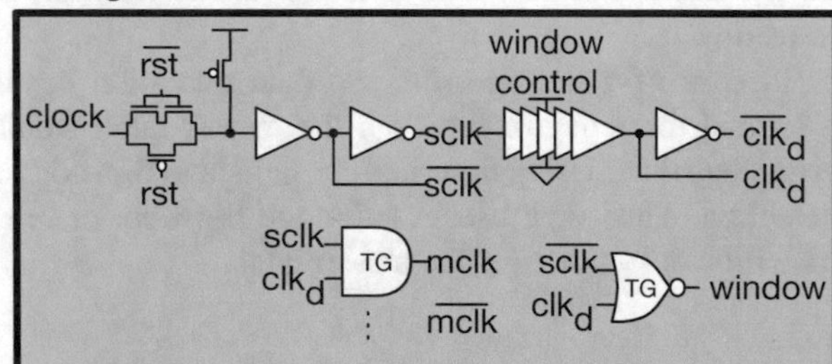

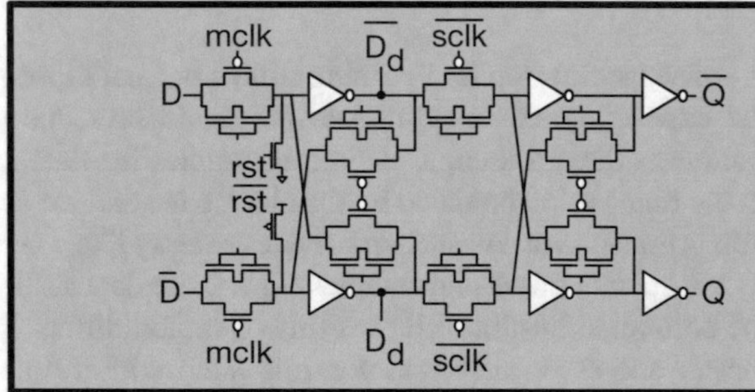

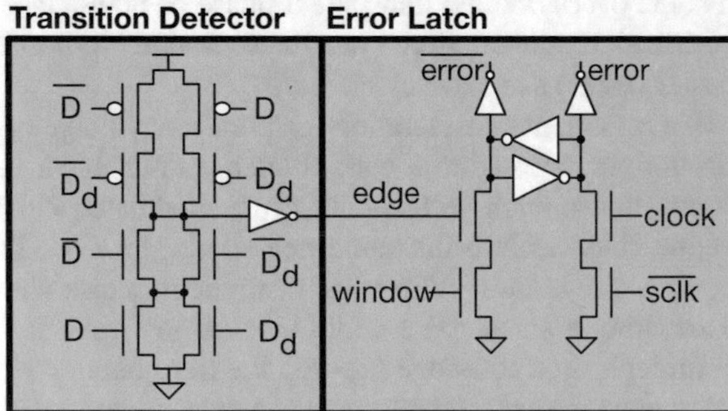

Fig. 6.2 Circuit of the proposed EDAC flip-flop

6.1.4 Error Latch

The transition detector flags any transition on the data input of the flip-flop. The error latch combines this data with the detection window to determine whether the flagged transition event is a timing error. In case of a timing error, the error signal should persist until the next clock cycle. Hence, a set-dominant latch (SDL) is used as the error latch. The latch can be enabled at any point during the detection window. Once set, the latch stays high until it is reset at the start of the next clock cycle. As such, any transition in the data during the detection window is considered a timing error, even when multiple transitions or glitches occur. Since the soft-edge flip-flop captures the correct data through time borrowing, the timing error flag stored in the error latch is not critical for correct operation. Hence, it can be synchronized at the next clock cycle. Additional error information processing occurs in the error processor.

6.1.5 Error Processor

The error processor combines the error information coming from all the EDAC-enabled flip-flops. Because the error latch synchronizes the $error$ signal, the error processor is able to do logic computation or processing on the errors. The complexity of the error processor heavily depends on the amount of $error$ signals and the scale of the system. The error processor equipped in this work combines $error$ signals prioritized according to the critical path they belong to. Additionally, it calculates a moving average of the errors per clock cycle and can trigger interrupts to the microcontroller depending on a programmable threshold. As such, the error processor makes the EDAC system error aware and can make error-informed DVS decisions to facilitate energy-efficient operation.

6.1.6 Timing Error Masking and Aware Operation

The combination of the five building blocks described in Sects. 6.1.1–6.1.5 enables timing error tolerant operation while scaling the voltage or frequency dynamically to enable energy efficiency. The soft-edge flip-flop results in inherent error correction, as data arriving in a deterministic timing window after the clock edge is still captured correctly. The transition detector flags these events and propagates them to the error processor through the error latch. This information is fundamental to the EDAC operation, as it gives information regarding the margined operation of the circuit. It enables safe reduction of the supply voltage below the worst V_{dd}. Such closed loop operation can facilitate near point-of-first-failure under a variety of conditions. When considering static (intra-die) variations, the supply voltage can be adapted to operate the system near-PoFF without post-fabrication calibration. When considering dynamic variations the closed loop system can track (relatively slow) varying conditions like temperature and supply voltage fluctuations.

A key difference with other methods like replica monitoring circuits is that EDAC operation allows operation at a better-than-worst-case V_{dd}, since the system is error resilient. Another key difference is that the monitoring occurs in situ, thus overcoming any remaining margin compensating for mismatch between the (replica) monitor and the actual critical path.

6.2 Circuitry

In Fig. 6.2 the transistor-level circuit of the EDAC flip-flop is shown. For the biggest part, the flip-flop is identical to the one equipped in Chaps. 2–4. It employs differential input–output signalling with a single-ended clock signal. The timing/control block creates a delayed clock using an inverter-based delay line. The delay line uses long gate length devices to maximize delay/area while minimizing leakage power. The $mclk$ and $window$ signal are created using transmission gate logic. The master latch is controlled using the delayed clock ($mclk$). The slave clock is identical to that of a normal flip-flop.

Apart from the data input nodes (D and $\overline{D}$), access to the intermediate nodes D_d and $\overline{D_d}$ is available. The transition detector uses all four signals in a complementary manner to realize an XNOR-like operation. Because of differential signalling, no additional logic is necessary to create these signals. Its truth table is shown in Table 6.1. In summary, it detects four different cases of transitions, represented

Table 6.1 Transition detector logic value truth table

D	D_d	$\overline{D}$	$\overline{D_d}$	$(D \vee D_d) \wedge (\overline{D} \vee \overline{D_d})$ → EDGE
0	0	0	0	0
0	0	0	1	0
0	0	1	0	0
0	0	1	1	0
0	1	0	0	0
0	1	0	1	1
0	1	1	0	1
0	1	1	1	1
1	0	0	0	0
1	0	0	1	1
1	0	1	0	1
1	0	1	1	1
1	1	0	0	0
1	1	0	1	1
1	1	1	0	1
1	1	1	1	1

Table 6.2 Overview of possible transition cases resulting in a transition detection

	Transition	Condition	Cause
Case 1	$D: 0 \rightarrow 1$	$\overline{D_d} = 1$	Master latch delay
Case 2	$\overline{D}: 0 \rightarrow 1$	$D_d = 1$	Master latch delay
Case 3	$D: 0 \rightarrow 1$	$\overline{D} = 1$	Diff. signal. mismatch
Case 4	$\overline{D}: 0 \rightarrow 1$	$D = 1$	Diff. signal. mismatch

in Table 6.2. Two are the result of the internal delay of the master latch, similar to a normal transition detector. The other two are the result of the differential signalling, enabled by the intermediary short in the pull-down network of the transition detector. This way, the transition detector can equally detect timing errors unique to differential signalling.

The error latch is set using the transition detector signal when the timing detection window is active. The cross-coupled inverter pair is enabled/reset using pull-down devices. Both during set and reset, the pull-down devices overpower the feedback in the latch to toggle the node. An error can occur at any point in the timing detection window. The latch is reset at the onset of the next detection window due to the overlap of the *clock* and $\overline{sclk}$ signals. This allows pipelined propagation of the error signals to the error processor as well as detection of new timing errors in the subsequent detection window.

6.2.1 Timing

In normal in-time operation, the timing error masking flip-flop behaves like a flip-flop. In timing-critical operation, the flip-flop facilitates error resilient operation. Detailed operation of the timing error masking flip-flop is shown in the timing diagram in Fig. 6.3. The timing/control block creates a delayed clock of which the master clock *mclk* and the timing window *window* are derived. This results in an overlap of the transparent phases of both latches, hence a hold time constraint. When data arrives in-time, as displayed in period T_{i-1}, operation is exactly like a normal flip-flop. The data propagation time in period T_i is larger than a full clock cycle, resulting in data arrival after the clock edge. In this case, the error detection flip-flop borrows time from period T_{i+1}. In this way, it provides error resilient operation. The data transition in the timing window triggers the transition detector. The transition detector in its turn triggers the error latch, setting the error signal high.

While the timing error masking flip-flop operates in conjunction with a normal flip-flop based pipeline and mostly behaves similar, it does have some altered timing constraints. The delayed master clock relaxes the setup time (t_{setup}) and tightens the hold time (t_{hold}) by allowing time borrowing. In Fig. 6.3 the flip-flop at the end of logic stage i is replaced by an error detection flip-flop. The constraints on the period T_i and T_{i+1} of logic stage i and $i + 1$, respectively, can then be described as in Eqs. 6.1 and 6.2.

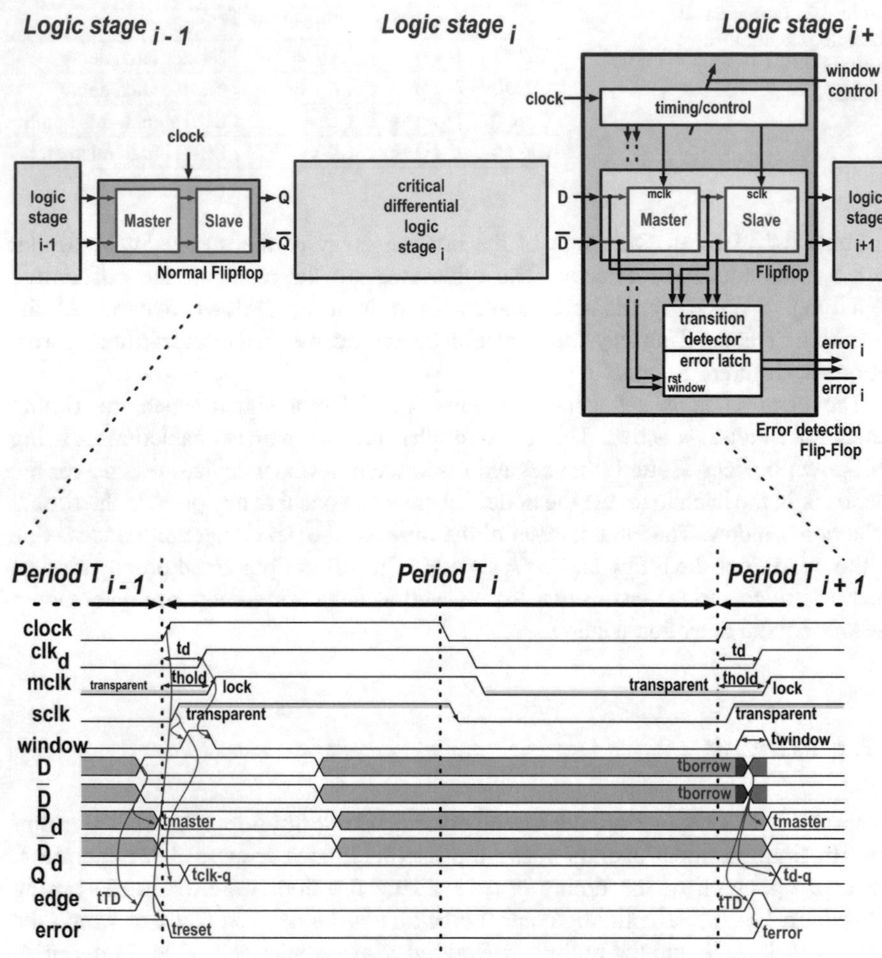

Fig. 6.3 Timing diagram of the proposed EDAC flip-flop

$$T_i \geq t_{\text{clk}-q,i-1} + \max(t_{p,\text{logic},i}) + t_{\text{setup},i} - t_{\text{borrow},i} \tag{6.1}$$

$$T_{i+1} \geq t_{d-q,i} + \max(t_{p,\text{logic},i+1}) + t_{\text{setup},i+1} + t_{\text{borrow},i} \tag{6.2}$$

$$\text{with } t_{\text{borrow}} \leq t_{\text{window}} \tag{6.3}$$

Both stages have to meet the setup constraint of their respective flip-flops. As the flip-flop launching the data in stage $i + 1$ is transparent, the hold time constraint for path i is tightened and can be described as in Eq. 6.4.

$$t_{\text{hold},i} \leq t_{c,\text{clk}-q,i-1} + \min(t_{p,\text{logic},i}) - t_{\text{window}} \tag{6.4}$$

As the flip-flop launching the data in stage $i+1$ is transparent, the data $- q$ propagation delay is part of the constraint on logic stage $i+1$ rather than the clock $- q$ delay in a normal flip-flop logic stage. While time borrowing relaxes the constraint on logic stage i, logic stage $i+1$ receives an additional penalty because of it. Note the difference between Eqs. 6.1 and 6.4. While the enabled time borrowing window is not necessarily equipped fully, it does reflect in full in the tightened hold time constraint. The full timing window width is added to the hold time constraint, while the actual borrowed time constrains logic stage $i+1$.

In the circuit of Fig. 6.2, the setup time constraint origins from the difference in delay between the clock and the data-path propagation for the master latch. Since the master latch clock edge is deliberately postponed, relating the setup time to the clock net in Fig. 6.3 results in a negative setup time. This corresponds to the intuitive analysis in Sect. 6.1: the data can arrive after the clock edge without corrupting the system.

6.2.2 ULV Implementation

The logic operation of the error detection flip-flop as described in the previous section results in timing error detection and resilient operation. Enabling this operation at ultra-low voltage poses a challenge, as some parts of the circuit are timing critical. The timing/control block relies on a delay line to create the derived clock signals. The most straightforward implementation of such a delay line is a chain of inverters. To achieve a good trade-off between area and leakage power overhead while creating enough delay to facilitate time borrowing, long gate length devices (60 nm) were used.

To minimize leakage power, the same long gate length devices were used to implement the flip-flop master and slave latch. This results in an increase of the master latch $d - q$ delay, of which the transition detector can benefit. To facilitate robust transition detection, the 1–1 overlap in $D - \overline{D_d}$ or $\overline{D} - D_d$ should be as long as possible. Increasing this delay tightens the constraint shown in Eq. 6.2. A better option is to improve the speed of the transition detector itself. This enables fast transition detection despite a very short 1–1 overlap. The complementary structure of the transition detector as shown in Fig. 6.2 aids to this, as well as the relatively strong pull-down devices in the applied 40 nm CMOS technology at ultra-low voltage.

The same holds for the set-dominant error latch: the pull-down devices toggling the latch have to be able to overpower the pull-up device that keeps the latch node high with a short pulsed signal coming from the transition detector. To this end, an inverter was added between the transition detector and the error latch. This enables strong pull-down to set the error latch as fast as possible rather than doing the same with weaker pull-up devices. In this set-dominant latch, the relative timing of the $edge$ and $window$ signal results in a setup time constraint. Even more crucial is the reset operation. In this work, it relies on the 1–1 overlap of the $clock$ and $\overline{sclk}$

signals. The latch is analysed and sized accordingly for correct set/reset operation at near-MEP supply voltages in the [0.2. . .0.5 V] range across process corners. Due to the significant change in pMOS/nMOS device strength between nominal and ultra-low voltage, the sizing of the error latch and its reset operation is what limits full voltage range operation.

6.2.3 Results

The timing error flip-flop allows a longer path propagation time while still providing correct operation. As mentioned in Sect. 6.2.1, this error resilient operation results in a negative setup time. Table 6.3 shows this setup time in comparison with that of a normal flip-flop across process corners. The difference (Δt_{setup}) directly represents the resiliency. It determines the amount of time data can arrive late in an EDAC flip-flop. Since the absolute time varies significantly between process corners, it is much more useful to relate these times to the clock period of the EDAC-enabled system. Since the system is equally sensitive to process variations, the system clock period scales accordingly. This scaling results in a more or less fixed ratio of additional data arrival time vs. clock period. This is discussed further in Sect. 6.4.2.1.

The timing error flip-flop introduces some overhead. Both an area and energy increase can be expected at the cell level. Table 6.4 compares the normal differential flip-flop equipped earlier in this work with the just presented error resilient flip-flop under typical conditions. Area, delay, leakage power and clock energy are increased. Despite this overhead, a key aspect of the system remains that the error resiliency it provides should reduce overall energy consumption. The timing/control block equipped in every timing error flip-flop results in the biggest area overhead. The transition detector and error latch are implemented with relatively low area. The area overhead due to timing error detection when comparing to the differential flip-flop implementation of Chap. 2 is 76–93%. As the flip-flop drive strength varies, the

Table 6.3 t_{setup} for normal flip-flop vs. EDAC flip-flop at 300 mV supply voltage

Process corner	Normal flip-flop t_{setup}	EDAC flip-flop t_{setup}	Δt_{setup}
TT	−0.80 ns	−8.32 ns	7.52 ns
SS	−5.55 ns	−50.93 ns	45.38 ns
FF	−0.19 ns	−1.83 ns	1.64 ns

Table 6.4 Normal flip-flop vs. EDAC flip-flop comparison under typical conditions at 300 mV supply voltage

	Normal flip-flop	EDAC flip-flop	Δ
Area [μm^2]	31.75	61.39	+93%
t_{setup}[ns]	−0.80	−8.32	−7.52 ns
$t_{\text{clk}-q}$[ns]	4.63	5.52	+19%
P_{leak}[nW]	21.79	41.32	+90%
E_{clk}[fJ]	2.34	4.12	+76%

size of the output buffer takes a larger relative area which results in a smaller relative overhead due to error detection. The total impact of such area overhead should be outweighed against the total area increase of the error resilient system, as well as the energy reduction it facilitates. While flip-flops are relatively large gates, the system area impact of the proposed EDAC flip-flop is small.

$t_{\text{clk}} - q$ is increase d by 19%. Its main cause is the additional loading on the master latch input/output nodes because of the transition detector. Due to the delay line in the timing/control block, clock energy is increased by 76%. Cell leakage power is increased accordingly by 52–90%. Sharing the timing/control block across multiple flip-flops similar to [15] can significantly reduce this overhead. However, such a strategy can compromise ULV operation since the constraints on the slew rate and relative skew of the $clock$ and $window$ signals are strict. As is shown in Sect. 6.4, the total area overhead attributed to timing error detection and timing error processing is limited to 7% due to sparse flip-flop replacement.

6.3 Modelling and Design Flow

An automated design flow to equip timing error detection in a digital system is necessary: if large scale integration on the same scale as current digital designs is impossible, the proposed technique is unusable. Such an automated design flow equips standard cell libraries with logic/timing/power information to facilitate synthesis, timing analysis and physical simulation and implementation.

6.3.1 Standard Cell Description

The complex behaviour of the timing error flip-flop is modelled to fit a standard cell description and is characterized across multiple voltages and corners. The functional descriptions used for the timing error flip-flop are shown in Fig. 6.4. A functional subdivision is chosen as to avoid output-to-output relationships and temporal dependencies. To this end, the $window$ signal is modelled as a logic level resulting from $sclk$ and $clkd$ rather than a pulse originating from the $clock$ signal. Intermediate signals (e.g., D_d, $edge$, $window$, $\ldots$) are characterized in a small range of slew/load conditions to accurately model the limited interconnect, slew and load at those nodes.

The combination of these three cells results in the operation as described in Sect. 6.2.1. The Synopsys SiliconSmart characterization tool is used to model the cells as any other. Other than propagation time and power consumption, flip-flop specific parameters, such as setup time, hold time, minimum pulse width, minimum period, etc., are characterized for different input slew rates and output loads. To accurately represent the time borrowing functionality of the system,

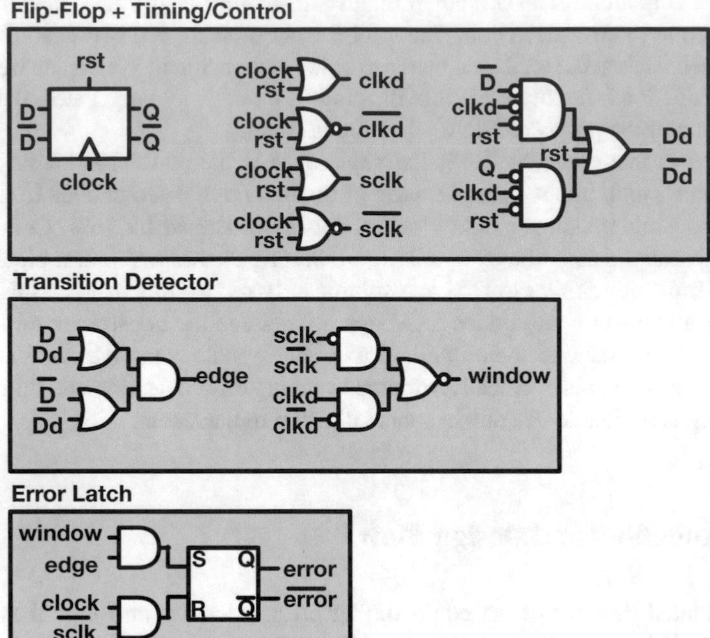

Fig. 6.4 Functional description of the timing error flip-flop used for characterization

the characterization tool models time borrowing as a negative setup time while significantly increasing hold time by an amount equal to the transparency window.

6.3.2 Augmented Design Flow

Section 6.3.1 accurately modelled the timing error flip-flop as a standard cell. While the model is accurate, logic synthesis tools cannot equip timing error flip-flops for their destined functionality. It remains up to the designer to insert the timing error flip-flops in the design. Despite the lack of synthesis integration, this work provides a timing error detection insertion strategy almost completely transparent to the designer. The outline of the design flow to realize an error detection enabled system is shown in Fig. 6.5. The design flow is, for the largest part, similar to the standard design flow. Replacement of normal flip-flops by timing error flip-flops occurs in the gate level netlist. While in theory this is possible after synthesis, replacement after clock tree synthesis and place-and-route yields better results. CTS and place-and-route have such impact on timing closure that omitting their effect when replacing the flip-flops can result in sub-optimal conclusions. Timing error flip-flop insertion occurs in the following steps:

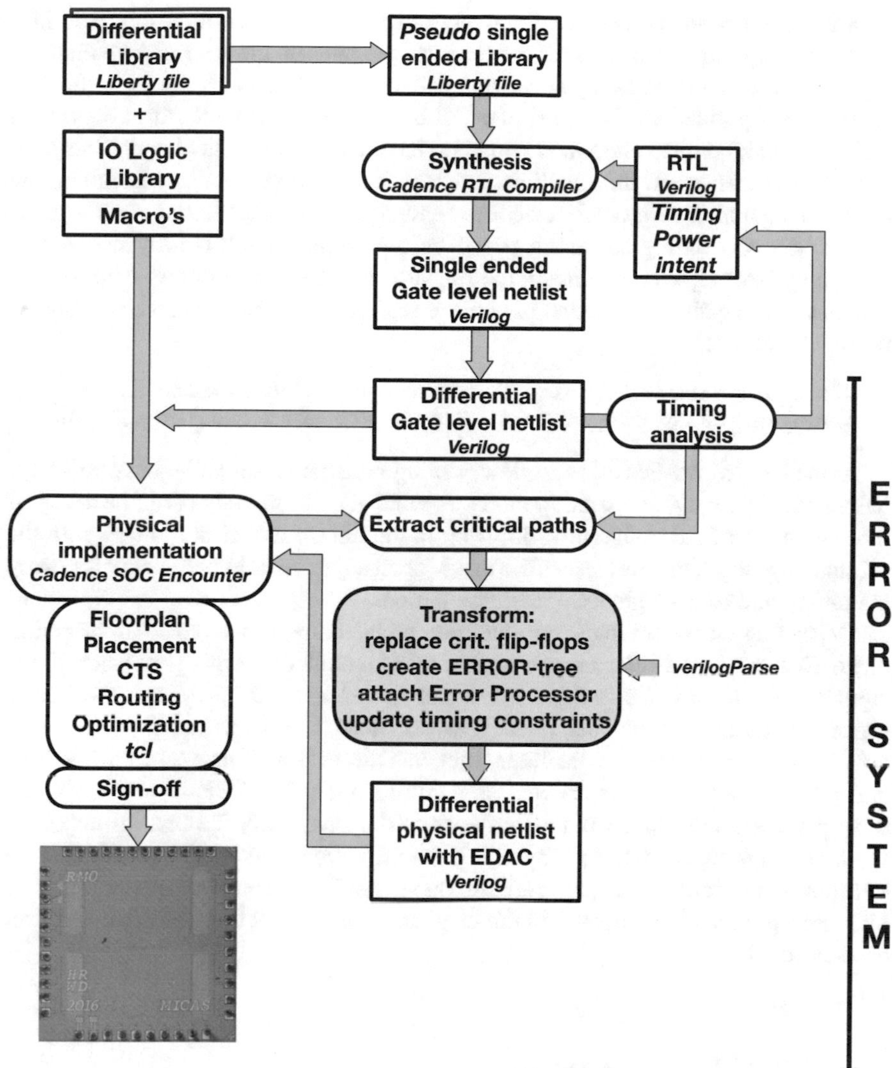

Fig. 6.5 Flow chart of the differential standard cell design flow augmented with timing error detection insertion

1. Run normal timing driven synthesis, placement, CTS, and routing.
2. Extract n most timing-critical endpoint flip-flops and slack.
3. Replace n most timing-critical endpoint flip-flops by timing error flip-flops.
4. Constrain critical timing paths according to original slack from (2).
5. Analyse timing and confirm timing-critical endpoints.
6. Re-iterate (1)–(5) if necessary.
7. Hold time optimization and finalize design.

Step 4 and 5 are crucial for the design. The timing error flip-flops exhibit a negative setup time which relaxes the timing constraint on the respective path. Step 4 prevents the tool from using the newly available slack due to timing error flip-flop insertion to optimize the design further. This aids step 5, in which it is confirmed that the original critical endpoints remained the most timing-critical endpoints after flip-flop replacement. If this condition is not met, other timing paths will fail before the paths equipped with error detection, rendering the EDAC technique unusable. Imposing a new timing constraint based on the original slack before insertion of the timing error flip-flop is done in the physical implementation flow. After timing analysis and selecting the number of paths to replace, these paths receive an adjusted timing constraint:

```
> set_max_delay old_delay_i -from * -to edacflipflop_i/D
> set_max_delay old_delay_i -from * -to edacflipflop_i/D_BAR
```

In the applied MMMC design flow, timing constraints on paths have to be set independently for every operating mode. A more generic strategy could be to adjust the setup time of the timing error flip-flop in the library model file in line with the original flip-flop. This overcomes the need for an adjustment in path delay for every operating mode during physical implementation.

Depending on the timing error detection architecture, other steps can be put in place. The microcontroller system in this work includes an error processor. *Error* signals are gathered at every hierarchy and routed to a single system level error processor, capable of making intelligent decisions based on the flagged timing errors. Additional steps to facilitate this include netlist adjustment to combine and prioritize *error* signals, as well as control signals for DVS. In this work, the error processor is included in the netlist from the start. Only the final *error* signal interconnect is adjusted during physical implementation iteration. Finally, hold time optimization is done. Short path padding due to hold time constraints is expected to increase significantly compared to the original design, as well as the area overhead associated with it.

6.3.3 Gate Level Simulation

The described standard cell representation and design flow augmentation result in a gate level netlist equipped with timing error detection and correction. The gate level representation of the timing error flip-flop is fully functional as described and allows timing-annotated simulation of the error detection and correction functionality. As such, the EDAC functionality of the implemented digital system can be verified.

Simulation of the EDAC functionality is realized by scaling the clock period rather than the supply voltage. The nominal clock period results in correct functionality without timing errors. Simulation with a critical clock period shows a timing error at the (data-dependent) critical path while still computing the correct result. Simulation with a sub-critical clock period compromises correct functionality.

The gate level simulation relies on static timing analysis to execute a timing-annotated simulation. Without considering data-dependent path activation, the timing errors realized in such a way always occur at the most critical path. As such, the gate level simulation confirms the static timing analysis results.

6.4 Timing Error Masking-Aware Microcontroller

6.4.1 Overview

To demonstrate the operation of the timing error flip-flop presented in Sect. 6.1 and evaluate the system level implications of such a timing error detection system, the architecture is integrated in a 32-bit microcontroller system. An overview of the system is shown in Fig. 6.6. Similar to the system presented in Chap. 4, it equips an ARM Cortex-M0 core with AHB-enabled peripherals (UART, GPIO, and TEST/DEBUG) and a 64 KB SRAM memory. Additionally, it includes an error processor and is equipped to feed back error information to the PCB level control loop. The system is realized using the design flow discussed in Sect. 6.3 and is

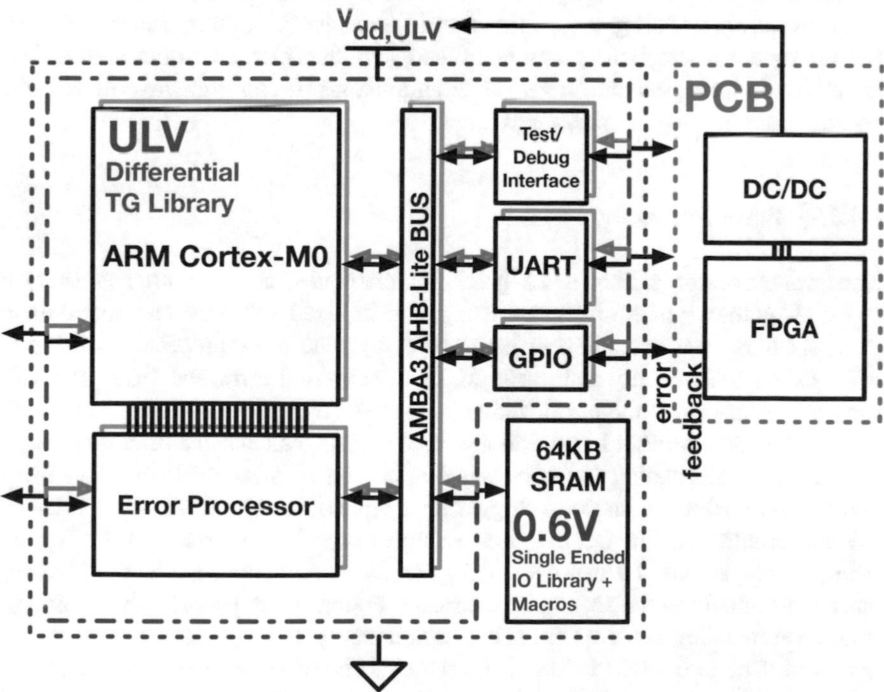

Fig. 6.6 System overview of the microcontroller system equipped with timing error detection

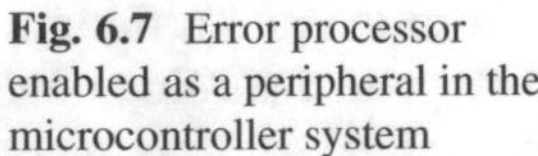

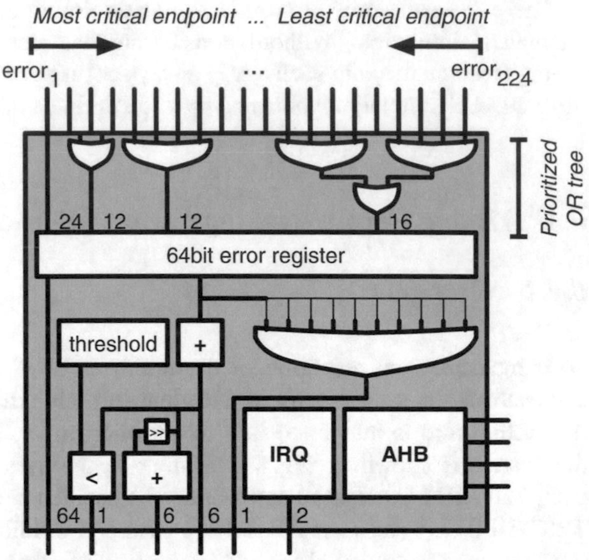

Fig. 6.7 Error processor enabled as a peripheral in the microcontroller system

transformed to enable timing error detection. This implies a sparse replacement of normal flip-flops by timing error detection flip-flops and an AHB-compatible error processor rendering the system timing error aware. Correct functionality of the timing error detection system combined with the microcontroller architecture is verified in timing-annotated gate level simulations. It confirms the critical paths coming from the static timing analysis.

6.4.1.1 Error Processor

The error processor, as shown in Fig. 6.7, is a distributed error capture and decision block. It gathers all error signals using a prioritized OR tree and maps them to a 64-bit register according to path criticality. The error processor is a trade-off between functionality and overhead. In its proposed form, the error processor provides access to all the prioritized error signals (64), provides comparison with a programmable threshold, sums all the error signals, calculates a running average and provides information on the presence of any error. Additionally, each of these signals can enable an interrupt, triggering a specific interrupt service routine in the microcontroller. The OR tree uses a single cycle; hence, errors can be flagged using interrupts within a two cycle delay. In its current form, the error processor adds substantially to the sequential overhead. Fixing the threshold value at design time or eliminating some of the debug functionality can significantly reduce this overhead. Due to the integration of the error processor within the architecture, the microcontroller can use dedicated subroutines according to error occurrence and control dynamic voltage scaling.

6.4.2 Design Trade-Offs

Several design trade-offs were made while implementing the EDAC microcontroller system. The most important considerations are the size of the detection window, and the critical path analysis with adjoining sparse flip-flop replacement.

6.4.2.1 Detection Window Selection

Applying an efficient detection window is imperative to balance energy and area overhead due to timing error detection compared to margins. The choice of detection window size in this design is based on four major design considerations:

1. The dynamic voltage scaling step size: During DVS the system should evolve from safe operation (zero errors) to some errors before corrupting the pipeline. Figure 6.8 illustrates this. In case 1 a data signal arrives early which results in correct operation, undetected by the error detection. This is typical for a relatively high V_{dd}. In case 2, the V_{dd} is reduced. The data propagation delay scales accordingly, to a point where the data arrives in the detection window. This results in a detected and corrected error. The turnover point between these two cases is the point-of-first-failure. Case 3 is the result of a large V_{dd} step. The data arrives outside the detection window and is indistinguishable from early data arrival. Such a case corrupts the pipeline due to a non-recoverable timing error. Here, the trade-off in detection window size is thus dynamic voltage scaling step size in comparison with a larger detection window. A small detection window results in very fine grained DVS susceptible to noise and difficult correct operation close to the PoFF.
2. The detection window directly determines the allowed amount of time borrowing. As all detected errors should also be corrected, the available time borrowing is identical to the detection window. More time borrowing can overcome more timing errors, but also tightens the constraint on the subsequent pipeline stage.

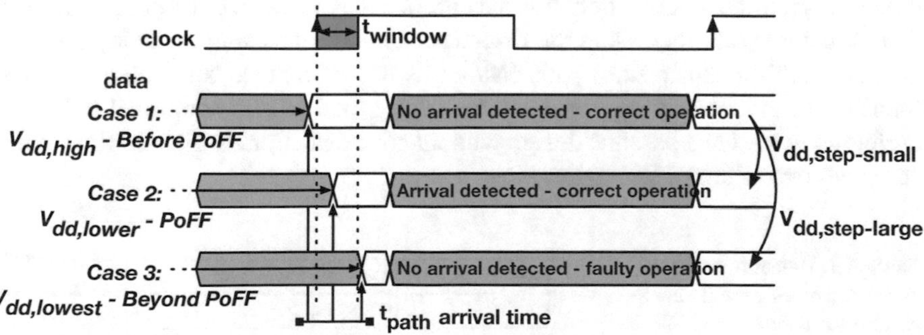

Fig. 6.8 Error detection as a result of DVS step size and detection window

This makes inherent error correction less accessible, or requires more timing error detection flip-flops to be equipped at paths with more slack.

3. The detection window is created in every flip-flop. While this is beneficial for skew between timing related signals like *mclk*, *sclk* and *window*, it introduces overhead in every flip-flop. A larger detection window requires more hardware overhead to be created. Table 6.4 shows the error detection related overhead in a single flip-flop. The delay line to create the detection window results in an area and leakage power increase. Externally biasing the delay line at a lower supply voltage can reduce the adherent overhead, but results in an additional supply voltage. As shown in Fig. 6.2 the delay line employed in this work is biased externally. This allows to scale the size of the detection window irrespective of the system supply voltage. As shown in the measurements reported later in this chapter, full functionality was achieved without the use of this external bias.
4. As shown in Eq. 6.4, the detection window directly impacts the hold time constraint. While timing error detection flip-flops are only equipped on critical path endpoints, an arbitrary number of short paths can have these flip-flops as endpoints. This results in a significant short path padding overhead. While the building blocks represented in this work prove to be variation resilient, hold time optimization is approached conservatively. This means a worst case (slow) condition is assumed for the clock capturing the data, while a best case (fast) condition is assumed for the clock and data-path launching the data. Under these conditions, the short path propagates as fast as can be, while the subsequent flip-flop locks this value as late as possible. Under ULV variation-prone conditions, extensive short path padding is necessary to overcome this hold time problem. Timing error detection through time borrowing enlarges the hold time, resulting in even more short path padding.

The presented 32-bit microcontroller system is equipped with the flip-flops as shown in the circuit in Fig. 6.2. The resulting detection window is shown in Table 6.3. Table 6.5 shows the resulting relative size compared to the system clock period under process variations. More than 5% of the clock period is guaranteed to be monitored using the 5 element delay line as shown in Fig. 6.2. Using the external bias voltage, a detection window of up to 25% can be achieved. However, hold time optimization was done for the nominal bias voltage, covering the cases as shown in Table 6.5. Increasing the detection window size using the delay line bias is thus optimistic, since short path padding is not guaranteed for such conditions. Short path padding due to error detection adds a total of 705 additional hold time buffers compared to a baseline design without error detection. Each hold time buffer combines two regular V_T inverters.

Table 6.5 Detection window vs. system clock period under process variations at 300 mV

Process corner	Detection window	T_{system}	%
TT	7.52 ns	117.48 ns	6.4%
SS	45.38 ns	584.92 ns	5.29%
FF	1.64 ns	30.95 ns	7.76%

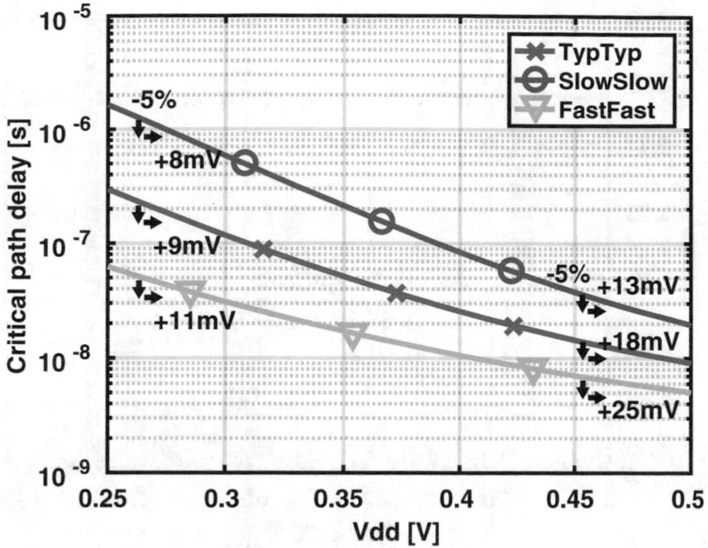

Fig. 6.9 T_{system} as a function of supply voltage for different process corners

To operate the system in a dynamic voltage scaling control loop, the supply voltage step size should be in accordance with the detection window size. Figure 6.9 shows the simulated speed of the critical path of the M0 system. In the entire voltage range, a 5% performance variation results in a minimum supply voltage variation of 8 mV. Such a step size is reasonable for the measurement setup as shown in Fig. 4.18. In an on-chip power management unit, the DVS step size should be taken into account when considering the overall system trade-offs.

6.4.2.2 Critical Path Analysis

The energy gains realized by timing error detection systems as described in Chap. 5 are possible largely due to an imbalanced timing histogram: while critical paths determine the maximum clock frequency, they are outnumbered by many non-critical paths. Timing error detection systems benefit from this property through sparse flip-flop replacement. Only the most critical path endpoint flip-flops are replaced. Since the timing error flip-flop introduces a significant overhead in power, area and short path padding, this allows timing error detection with a limited overhead.

Figure 6.10 shows a histogram of all the endpoint flip-flops ordered according to the smallest timing slack (longest delay) path they serve. The amount of replaced path endpoint flip-flops maps to the amount of timing slack 'covered' by timing error detection. A replacement of the 224 most critical endpoint flip-flops results in a monitoring range of 15% of the clock period. This means any timing path

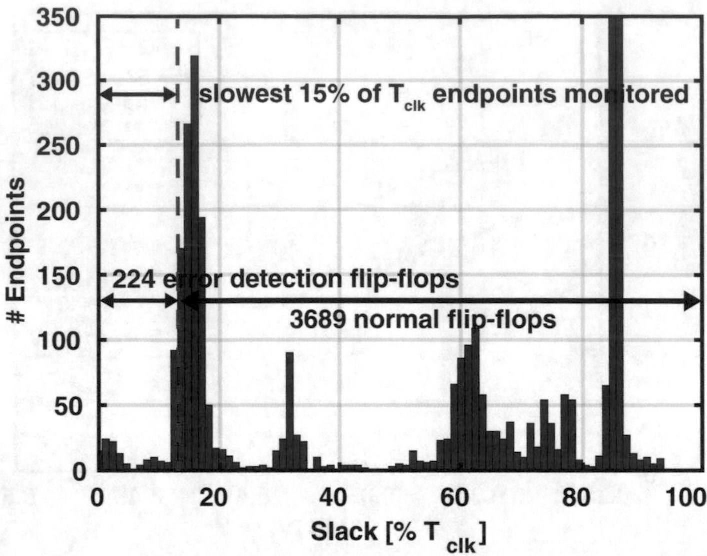

Fig. 6.10 Histogram of the path with the smallest timing slack at each endpoint flip-flop

with a propagation delay larger than 85% of the clock period is equipped with a timing error flip-flop. 3689 normal flip-flops remain in place, which amount to a replacement rate of 5.7%. The decision to replace exactly this amount of flip-flops is threefold.

First, the histogram shows a large number of endpoints in the 15–20% slack range. As timing error flip-flops introduce an area and power overhead, a large number of these flip-flops would offset the energy consumption too much. The 932 endpoints in this slack range would result in a replacement ratio of 30%. It is clear such a power and area overhead would be hard to overcome by eliminating margins and operating at the PoFF.

Second, fail-safe operation of the circuit should be guaranteed. From static timing analysis, the selected 224 endpoints are the most critical and will fail, by definition, first. Several factors such as voltage and temperature delay dependence, data-dependent path activation and process variations can change this perspective. In this work we focus on the influence of process variations. Fail-safe operation is guaranteed when, under any process variation condition, one or more timing error detection enabled paths fail (detect an error) before a single normal path does. In such a consideration, the delay of every timing path in the histogram in Fig. 6.10 can be represented as a (log-)normal distribution of which the mean and standard deviation arise from process variations. Guaranteeing fail-safe operation then means making sure the chance of such failure is very small. This chance can be calculated as follows:

$$P_{\text{false pos}} = P(\exists\, p_i : t_{\text{prop},p_i} > \max(t_{\text{prop},q_j})) \tag{6.5}$$

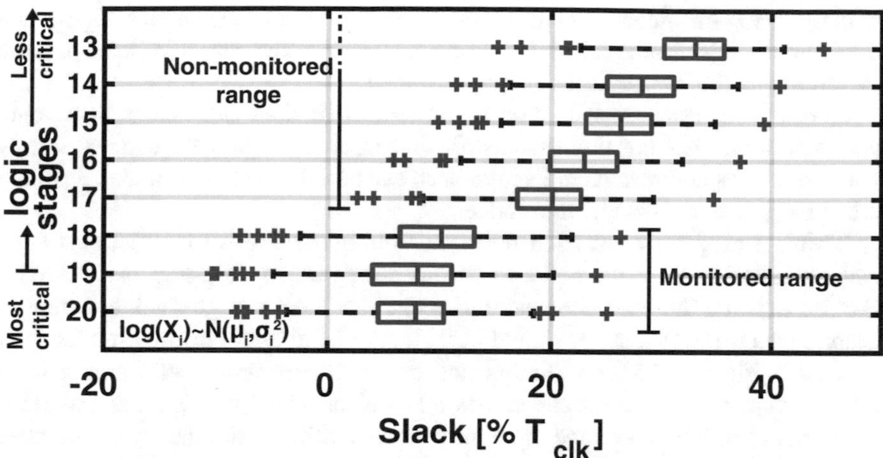

Fig. 6.11 Boxplot of the slack distribution of a subset of timing paths, acquired through 300MC simulations at 300 mV

with $p_{:}$ representing all non-monitored paths and $q_{:}$ all monitored paths. Because extracting the statistical data from each timing path is computationally intensive, the problem is simplified. The most critical path contains 20 logic stages. By redefining the bin histograms to a 5% covering range, 20 bins remain. Out of each bin a representative timing path is selected with the appropriate stage length. The log-normal distribution of each of these paths is determined using Monte Carlo simulations. The resulting distributions are shown in Fig. 6.11. The endpoints in the histogram are redistributed according to the 20 bins in the boxplot. The inequality in Eq. 6.5 simplifies to

$$P((t_{\text{prop},p_i})_{\text{any path in bin 1..17}} > (t_{\text{prop},q_j})_{\text{all paths in bin 18..20}}) > 0 \tag{6.6}$$

In calculating the difference between two log-normal distributions (logX − logY), it is easier to compare the ratio of the log-normal distribution of their respective normal distributions (log(X/Y)). With this in mind, the full chance of false positive events can be written as follows:

$$\begin{aligned} P_{\text{false pos}} = {} & (ep_{17}) * P(t_{p,20} < t_{p,17})^{ep_{20}} + (ep_{17}) * P(t_{p,19} < t_{p,17})^{ep_{19}} \\ & + (ep_{17}) * P(t_{p,18} < t_{p,17})^{ep_{18}} + (ep_{16}) * P(t_{p,20} < t_{p,16})^{ep_{20}} \\ & + (ep_{16}) * P(t_{p,19} < t_{p,16})^{ep_{19}} \quad + (ep_{16}) * P(t_{p,18} < t_{p,16})^{ep_{18}} \\ & + \cdots + (ep_{2}) * P(t_{p,20} < t_{p,2})^{ep_{20}} + (ep_{2}) * P(t_{p,19} < t_{p,2})^{ep_{19}} \\ & + (ep_{2}) * P(t_{p,18} < t_{p,2})^{ep_{18}} + (ep_{1}) * P(t_{p,20} < t_{p,1})^{ep_{20}} \\ & + (ep_{1}) * P(t_{p,19} < t_{p,1})^{ep_{19}} + (ep_{1}) * P(t_{p,18} < t_{p,1})^{ep_{18}} \end{aligned}$$

with ep_i being the amount of endpoints represented in bin i. Using the log-normal distributions, this chance calculates to be 2.3×10^{-17}. The chance of false positive monitoring in this system is thus negligibly small.

Note that this approach has some limitations. It does not take into account path activation rates nor timing redistribution in different process corners, supply voltages or temperatures. A similar approach can be used to do so, but would require much more data and system knowledge.

Third, timing error detection relies on time borrowing, so it constrains logic paths originating from the replaced timing error masking flip-flop (see Eq. 6.2). For this to work, those attached paths should either have enough slack to complete despite the borrowed time or should in their turn be able to borrow time for their next stage. Figure 6.12 shows timing statistics after replacement of the mentioned 224 flip-flops across process corners. Non-EDAC attached paths have at least 15% slack available, but on average often have close to 50% slack remaining. Attached paths with EDAC-enabled flip-flops have much less slack (at least 3%), as can be expected, but can borrow time from their adjacent paths. These path endpoints return in the original 224 flip-flops, making the analysis recursive. Non-EDAC attached paths thus provide sufficient slack for time borrowing to occur and still complete correctly.

6.4.3 SRAM Interface

Since the SRAM used in this work is a commercial macroblock, the memory input interface is not capable of detecting timing errors similar to the timing error detection flip-flops. Timing critical memory accesses can thus not be overcome. Some work has been published on detecting timing errors for memories [7], often relying on replica bits. The bitcell macro of an SRAM allows few adjustments, thus SRAM EDAC strategies focus on the readout circuit. Additionally, a speculative write operation often comes at a great cost. Write operation is thus margined carefully, and EDAC operation only focuses on speculative read operation. In this work, an additional pipeline stage is inserted at the SRAM interface to avoid critical paths at the input of the SRAM. This results in a single cycle delay penalty during SRAM access but overcomes the need for an EDAC-enabled SRAM macro.

6.5 Measurements

The system as described in Fig. 6.6 was fabricated in a 40 nm General Purpose CMOS process. The die micrograph is shown in Fig. 6.13. Active area is 0.41mm^2, of which 46% is memory. It instantiates 12,424 cells of which 3913 are flip-flops. The 224 timing error flip-flops are highlighted in the floor plan in Fig. 6.14. The baseline design presented previously allows a careful comparison in performance

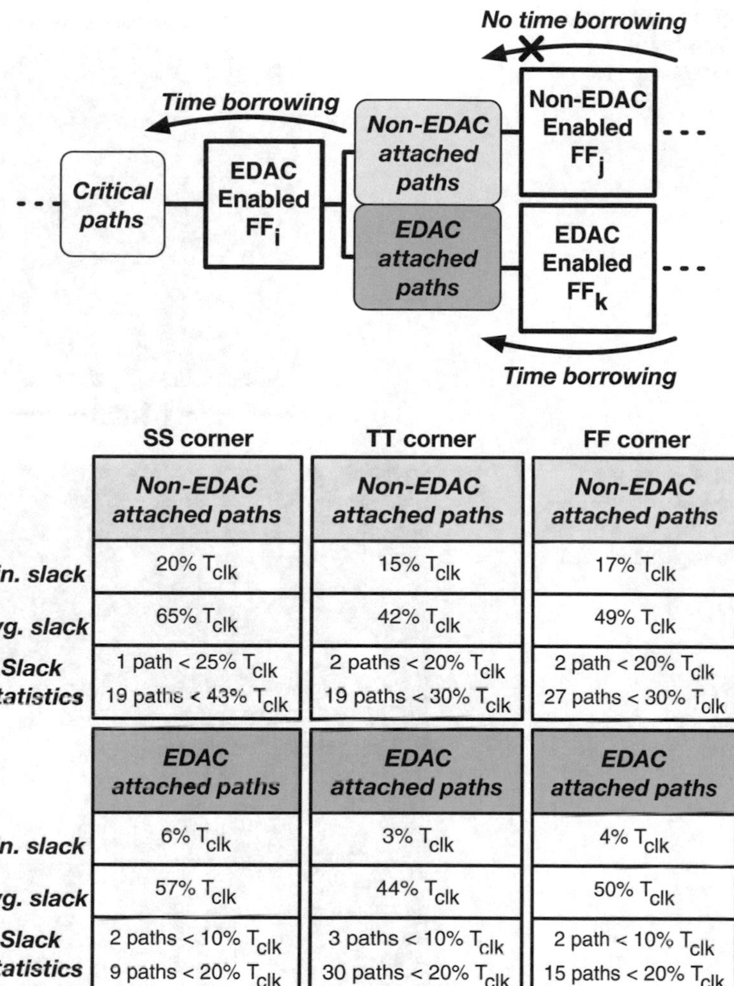

	SS corner	TT corner	FF corner
	Non-EDAC attached paths	*Non-EDAC attached paths*	*Non-EDAC attached paths*
Min. slack	20% T_{clk}	15% T_{clk}	17% T_{clk}
Avg. slack	65% T_{clk}	42% T_{clk}	49% T_{clk}
Slack Statistics	1 path < 25% T_{clk} 19 paths < 43% T_{clk}	2 paths < 20% T_{clk} 19 paths < 30% T_{clk}	2 path < 20% T_{clk} 27 paths < 30% T_{clk}
	EDAC attached paths	*EDAC attached paths*	*EDAC attached paths*
Min. slack	6% T_{clk}	3% T_{clk}	4% T_{clk}
Avg. slack	57% T_{clk}	44% T_{clk}	50% T_{clk}
Slack Statistics	2 paths < 10% T_{clk} 9 paths < 20% T_{clk} 21 paths < 30% T_{clk}	3 paths < 10% T_{clk} 30 paths < 20% T_{clk} 51 paths < 30% T_{clk}	2 path < 10% T_{clk} 15 paths < 20% T_{clk} 47 paths < 30% T_{clk}

Fig. 6.12 Slack analysis of EDAC equipped paths vs. non-EDAC equipped paths

and overhead, since the RTL code is identical except for the error detection system. A total of 7% of the active area was necessary to enable error detection and processing.

The chip is operated in a PC-controlled measurement board equipped with clock control, dynamic voltage scaling and power measurement (see Fig. 4.18). All I/O signals are accessible through an FTDI interface. The on-chip SRAM memory is loaded with the boot sequence, the instruction memory and the data memory using the debugger before disabling the reset signal. An off-chip generated pulse train acts as the clock. This allows single clock cycle level control of the system. C-code

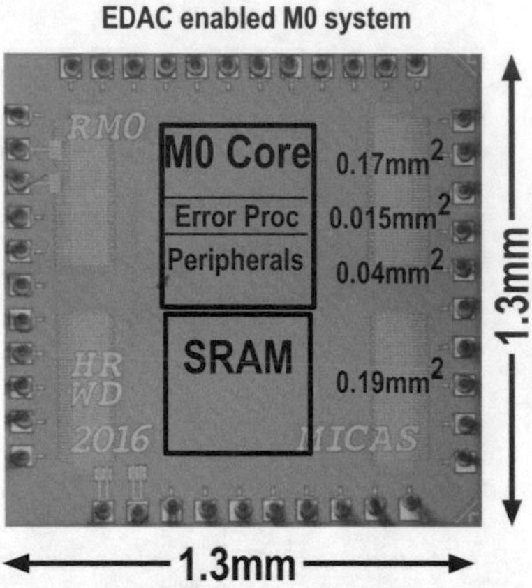

Fig. 6.13 Die micrograph of the EDAC-enabled microcontroller system

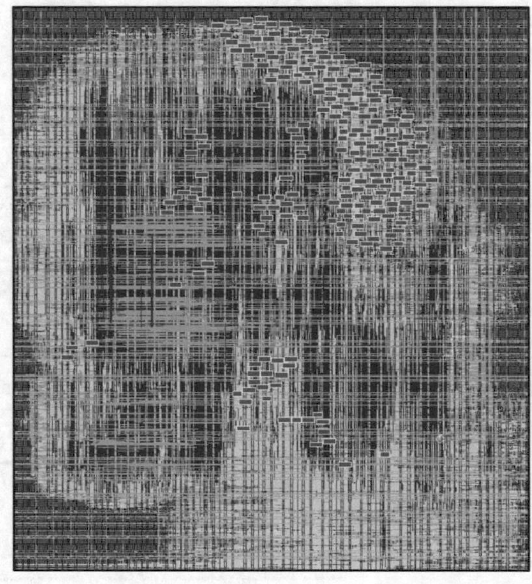

Fig. 6.14 Layout of the microcontroller system floor plan with 224 placed timing error flip-flops highlighted

compiled using the ARM tool chain can be run on the chip. For the measurements in the next sections, the Dhrystone benchmark C-code [13] has been used as representative code for the microcontroller operation.

6.5.1 Dynamic Voltage Scaling and PoFF Performance

The microcontroller system is fully functional and is operated with the PCB level DVS loop. Target frequencies between 5 and 30 MHz are set. At the predetermined target frequency, operation is started at a conservative voltage. The supply voltage

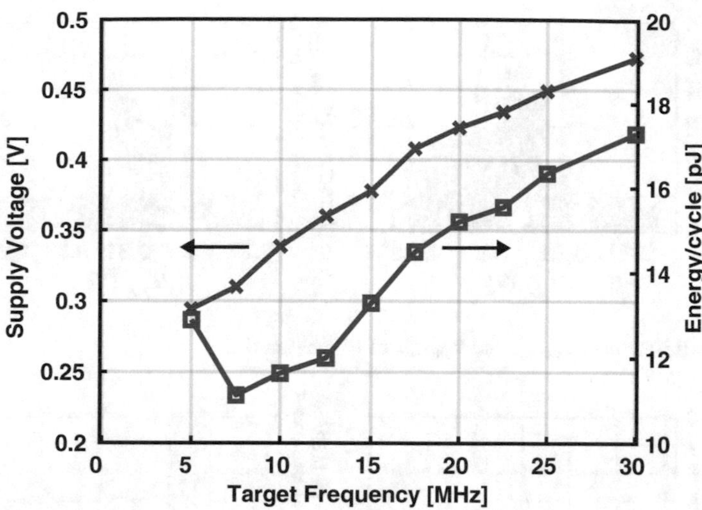

Fig. 6.15 Measurement of the PoFF curve for a wide frequency range, showing required supply voltage and energy consumption for the achieved performance

is gradually decreased. The error detection enables an I/O signal when a single error has occurred, indicating the point-of-first-failure has been reached. At this point, the control loop is ceased and correct operation is verified, while power consumption and leakage current are measured simultaneously. This measurement procedure is used to generate the PoFF measurement results in Fig. 6.15. It shows the required supply voltage and energy consumption to achieve the depicted range of target frequencies. Measurement results are averaged across 14 dice. Correct operation is achieved down to 290 mV. The minimum energy point is achieved at a slightly higher voltage and frequency: 7.5 MHz and 310 mV achieve 11.12 pJ/cycle energy consumption.

6.5.1.1 Variations

Across the measured dice, the minimum supply voltage and MEP supply voltage vary. Figure 6.16 shows both histograms. The change in supply voltage also has an effect on the energy consumption variation between different dice, as shown in Fig. 6.17.

As mentioned in Sect. 6.4.2.2, factors like process variation or temperature can influence which of the monitored critical paths fails. To shed light on the these variations, 10 MHz operation is compared between different dice and at different temperatures. Figure 6.18 shows the $V_{dd,PoFF}$, as well as the failure rate of the monitored paths, grouped as in the 64-bit register in the error processor depicted in Fig. 6.7. This measurement is done for 6 different dice. Every die has a different $V_{dd,PoFF}$ for 10 MHz operation, according to the timing errors it reports.

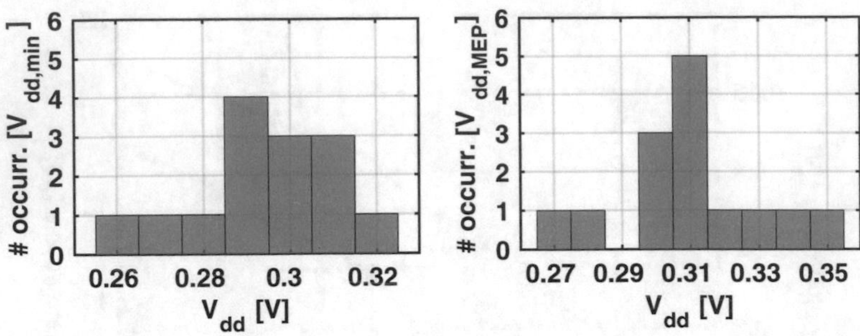

Fig. 6.16 Histogram of $V_{\mathrm{dd,min}}$ and $V_{\mathrm{dd,MEP}}$ of 14 measured dice

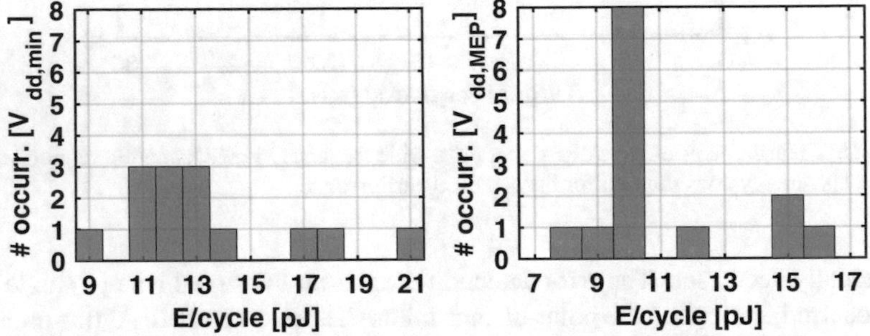

Fig. 6.17 Histogram of energy/cycle at $V_{\mathrm{dd,min}}$ and $V_{\mathrm{dd,MEP}}$ of 14 measured dice

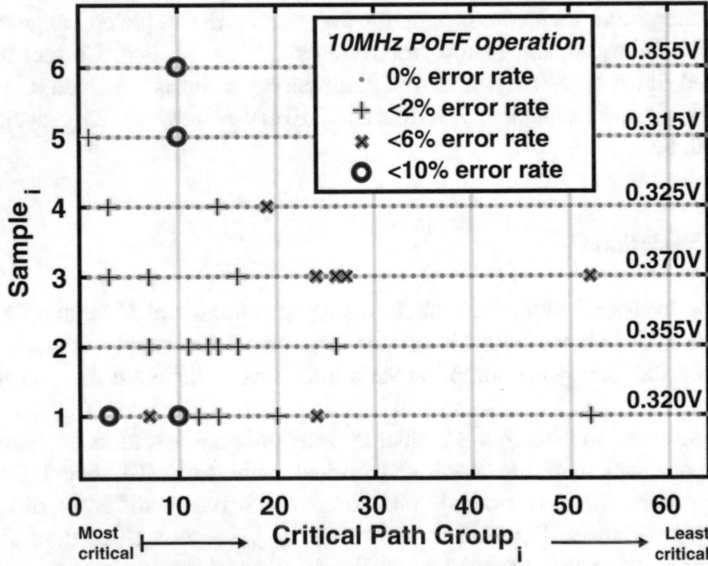

Fig. 6.18 PoFF supply voltage and average error rate of monitored path groups for 6 different dice

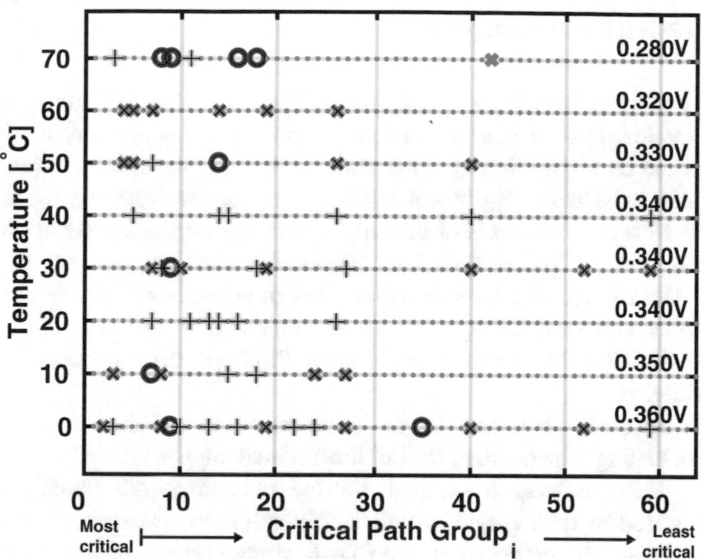

Fig. 6.19 PoFF supply voltage and average error rate of monitored path groups for a single die across a 0–70 °C temperature range

Additionally, the per-path failure rate is shown. The failure rate percentage is the result of the detected errors present in the error processor after a burst of 37 clock cycles. This readout is repeated 3600 times to average out the influence of path activation rate and supply noise. As can be seen in Fig. 6.18, most timing errors occur in the 27 most critical path groups, corresponding to the 30 most critical endpoint flip-flops. Error prone paths vary significantly between individual samples. As expected, intra-die variation results in a mismatch between static timing analysis critical path prediction and actual fabricated critical paths.

Figure 6.19 shows a similar analysis for a single die across a 0–70 °C temperature range. As in process variations, every temperature demonstrates its own $V_{dd,PoFF}$. A change in temperature again results in different critical paths. While most timing errors still occur in the 30 most critical endpoint flip-flops, less timing-critical paths fail as well. Path 40, for instance, does not produce timing errors at nominal temperature, but does fail more often at high or low temperatures. An important consideration here is the analysis occurs at equal frequency performance. The V_T of the devices shifts significantly as a result of the temperature shift. Process variations are exacerbated due to operation closer to the threshold voltage, which is reflected in the path failure rate. These measurements show the importance of intra-die variations influencing the sign-off strategy. The applied critical path analysis and sparse flip-flop replacement were quite conservative and thus overcome these effects. Taking into account the effects of temperature, voltage and process variation on EDAC timing path selection can improve the performance of such a system even further.

6.5.2 Baseline Comparison

Because the applied microcontroller system is identical to that of Chap. 4, it provides opportunity for detailed comparison. RTL code and functionality are identical, except for the error detection and processing. The baseline system is run at point-of-first-failure: the lowest V_{dd} at which the system operates correctly for each target frequency. New measurements of the baseline design were done at the same target frequencies as the EDAC design. This allows a detailed comparison, as shown in Fig. 6.20. On average, the EDAC-enabled design consumes 27–37% more energy to achieve the same target frequency. This overhead is due to the circuit overhead introduced by EDAC operation: timing error flip-flops, error processor and short path padding.

In this setup, the baseline design is measured in an ideal condition. Post-fabrication testing is performed to find the optimal supply voltage for each target frequency. Each die requires a new routine of testing. All on-chip hardware overhead added to such a system will result in energy overhead, as reflected in the measurement. By comparison, the EDAC system operation is the result of the autonomous control loop which decreases the supply voltage down to the PoFF. It requires no post-fabrication testing to achieve this performance, and every die achieves its own optimal supply voltage without intervention.

The described energy overhead comes with augmented functionality. It allows optimal energy performance of every die for the specified target frequency. As such, it also overcomes the poor predictability of near-threshold operation. This becomes clear when looking at the baseline simulation energy also shown in Fig. 6.20. In the upper left corner, the slow–slow sign-off points are shown: 1 MHz for a 350 mV

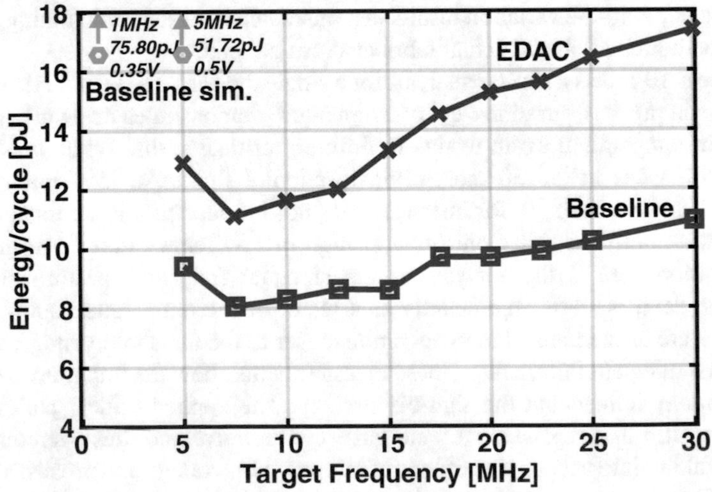

Fig. 6.20 Energy measurement of the PoFF curve for both the EDAC and the baseline design

supply voltage and 5 MHz for a 500 mV supply voltage. Operating the baseline chips at these operating points, the measured energy consumption is 75.80 and 51.72 pJ/cycle. Without post-fabrication testing, there is no guarantee that any of the fabricated baseline chips will function correctly at a faster target frequency for the quoted supply voltages. A more realistic comparison is thus to compare the slow–slow operating point of the baseline design to the EDAC design, as they provide the same level of guarantee regarding correct operation at the desired operating point. In this comparison, the energy consumption of the EDAC design is 75% lower than the baseline design.

6.5.3 *Replica Circuit Comparison*

The slow–slow corner operating points used in the previous section are pessimistic. Other techniques with a lower complexity than error detection and correction have been used to overcome these fairly bad operating points. Replica-based monitoring is one of those. In order to compare the EDAC system performance to replica circuit based prediction, the baseline design from Chap. 4 was equipped with an on-chip 21-stage ring oscillator. While the 21 stages do not directly mimic the critical path of the microcontroller system, they do experience close to the same PVT conditions. Using *SPICE* simulations, the ring oscillator speed can be mapped to the speed of the critical path of the microcontroller system. Figure 6.21 shows an example of

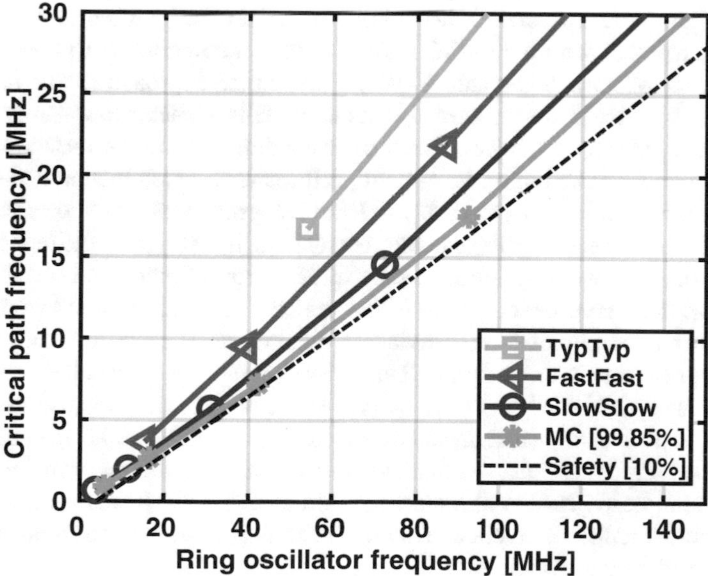

Fig. 6.21 Ring oscillator speed related to critical path speed at different process corners and under intra-die variations

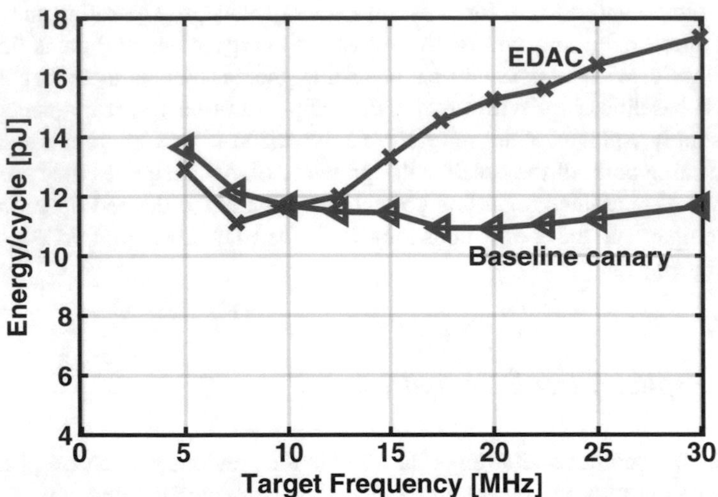

Fig. 6.22 Energy measurement of the PoFF curve for both the EDAC and the baseline design

how both these speeds relate. The figure also shows how a margin covering 99.85% of the intra-die (MC) variation influences their relationship, as well as an additional 10% safety margin.

An interpolated process corner of the every die is estimated using the relationship shown in Fig. 6.21. This interpolation is used to estimate the microcontroller system speed, using the critical path speed as a reference. The energy consumption of the baseline system operating according to the replica prediction is shown in Fig. 6.22, together with the EDAC system energy. Comparing both clearly shows the influence of intra-die variations. While the replica approach results in a lower energy at high frequencies (higher V_{dd}), it clearly fails to enable low energy at lower frequencies and supply voltages. Under these conditions, intra-die variation margins result in a poor prediction of the system performance, which increases the energy consumption. The in situ aspect of the EDAC approach clearly shows its benefit here. The margin necessary to guarantee correct functionality with the replica circuit is higher than the overhead introduced by timing error detection and correction.

An important consideration here is that the effort to predict circuit performance using the replica circuit is non-trivial. It requires post-fabrication measurement of the replica circuit on each die. This effort should be outweighed against the autonomous operation of the EDAC based system. Another important consideration is that the replica approach as depicted here assumes a perfect match in temperature, process corner, supply voltage, circuit ageing, etc. A more thorough analysis taking into account these possible differences would significantly increase the necessary margin on the replica prediction. This would shift the comparison in favour of the EDAC based system even further.

6.5.4 *Variation Resilience*

In this work, the focus is on process variations. The fact that they are fixed after fabrication favours some design considerations. A relatively small detection window is possible for static variations, as the DVS control loop starts from a conservative voltage. Some EDAC publications have shifted their focus on variations other than process variations. As mentioned in Sect. 5.5, temperature, ageing, jitter and voltage droop can equally produce timing errors. In theory, timing error detection can overcome any of those variations. Compared to process variations, these variations are much more dynamic and can thus produce more severe timing errors. In that context, the considerations regarding the detection window size as presented in Sect. 6.4.2.1 do not hold. A sudden voltage droop is much like lowering the supply voltage by $V_{\mathrm{dd,step}}$ in Fig. 6.8. A too large voltage droop could result in data arrival outside the detection window which results in system failure without error detection.

To overcome dynamic variations such as voltage droop, a more thorough analysis on their expected influence is necessary. Careful co-design of the EDAC system with traditional voltage droop mitigation strategies (e.g., decoupling capacitance) is needed. Simply increasing the detection window size has severe overall system impact, and could result in more energy and area overhead due to short path padding and detection window generation than simply adding more decoupling capacitance.

As discussed in Chap. 2, an ultra-low energy system operating at the MEP has a relatively high static energy. This holds for the presented prototype. As a result, the impact of a change in dynamic energy on the supply voltage is relatively small. A more thorough investigation evaluating the impact of voltage droop on the presented EDAC architecture is necessary to evaluate its properties for such applications.

6.6 State-of-the-Art Comparison

The comparisons in Sects. 6.5.2 and 6.5.3 are the most relevant, as they compare with the same system under different conditions. This section compares with several recent state-of-the-art EDAC implementations. Since implementations may differ significantly, a thorough comparison of the EDAC strategies is shown in Table 6.6. It compares with eight recent EDAC strategies implemented on (more or less) similar systems.

The upper half of each table compares the detection and correction strategies and how they impact clock generation, detection window, hold time constraints, area and ultra-low voltage operation. The impact of clock generation and detection window generation can skew the overall comparison heavily. All works rely on a limited detection window to capture erroneous data. This means they assume correct data to arrive within the detection window (thus $t_{\mathrm{prop,total}} \leq T_{\mathrm{clk}} + t_{\mathrm{DW}}$). The variation mitigation capability of all these works therefore primarily depends on the detection capability rather than the correction strategy. The most important consideration in

Table 6.6 Performance summary and state-of-the-art comparison of different EDAC implementations

	This work [12]	Razor II [3]	TDTB/DSTB [1]	DSTB [2]
Sequential element	Soft-edge flip-flop	Latch	Latch	Flip-flop
Extra transistors	46	31	15/26	28+delay chain
Detection	In-latch TD	TD	TD/DS	TD
Clock (duty cycle)	50%	13% and 40%	Controllable DC	50%
Detection window, DW	Local gen. 5% T_{clk}	Local 25FO4	High phase T_{clk}	Local
t_{hold} constraint	t_{DW}	t_{DW}	t_{DW}	$\sim t_{DW}$
Correction	Time borrow	Time borrow instr. replay	$T_{clk}/2$ instr. replay	$T_{clk}/2$ instr. replay
FF_{area} overhead	+76…92%	–	–	–
FF_{clock} overhead	+120…180%	+25…70%	+38…64%/ +81…143%	–
Near-V_T enabled	Yes	No	No	No
	0.29 V		0.7 V	0.9 V
Architecture	32-bit Cortex-M0	64-bit alpha	3-stage test circuit	6-stage ARM proc.
Technology	40 nm CMOS	130 nm CMOS	65 nm CMOS	65 nm CMOS
$System_{area}$ overhead	7%	–	–	6.9%
Sparse insertionrate	224/3913	121/826	–	503/2976
% #FFs	5.7%	14.6%		17%
Energy comparison	Margined/ unmargined baseline	Margined EDAC	Margined baseline	Sim. basel./ marg. EDAC
	−75%/+26…37%	−30…36%	−31…37%	+9.4%/−24%

	Bubble Razor [4]	RZL [14]	Razor-Lite [8]	Rproc [6]	iRazor [15]
Sequential element	Two-phase latch	Latch	Flip-flop	Two-phase latch	Latch
Extra transistors	20+dynOR+cluster	29	8	24	3+6.5[a]
Detection	DS	TD	Virtual V_{dd} TD	Virtual V_{dd} TD	Virtual V_{dd} TD
Clock (duty cycle)	2-phase non-overlap.	Controllable DC	Controllable DC	2-phase non-overlap.	*not given*
Detection window, DW	High phase T_{clk}	Global low phase T_{clk}	High phase T_{clk} post-fabr. calibr., 16%	High phase phase T_{clk}	~High phase T_{clk}
t_{hold} constraint	None	High phase T_{clk}	High phase T_{clk}	None	t_{DW}
Correction	Stall (bubble)	Time borrow interpolation	$T_{clk}/2$ instr. replay	V_{dd} boost	1-cycle stall
FF_{area} overhead	–	–	+33%	+268%	+4.3%
FF_{clock} overhead	+88%	+16.9%	–	–	–
Near-V_T enabled	No	No	No	Yes	No
	0.68 V	0.85 V	0.83 V	0.29 V	0.6 V
Architecture	32-bit Cortex-M3	16-bit FIR	64-bit alpha	16-bit R proc.	32-bit Cortex-R4
Technology	45 nm SOI	65 nm CMOS	45 nm SOI	65 nm CMOS	40 nm CMOS
$System_{area}$ overhead	87%	/	4.42%	8.3%	13.6%
Sparse insertion rate	–	118/393	492/2482	57/445	1115/12875
% #FFs	100%	30%	20%	13%	8.7%
Energy comparison	Margined EDAC	Margined EDAC	Margined EDAC	Margined baseline	Margined baseline
	−54...62%	+25...37% efficiency	−45.4%	−33...51%	−33...41%

[a]Due to shared local clock generation

variation resilience is whether or not the variation source generates timing errors within the detection window, and how fast this can change.

The lower part of each table focuses on the implemented system and its EDAC overhead. Different architectures can heavily influence the overhead, especially when considering short path padding and sparse flip-flop replacement. As with most EDAC implementations, our work keeps overall area overhead low due to sparse flip-flop replacement and large parts of non-sequential logic or macros (SRAM).

Different implementations report different energy reductions depending on the adopted baseline comparison. This work reports measured net energy increase compared to a separate baseline silicon implementation, optimized for operation without EDAC system. Few other works report energy compared to a separate silicon implementation; most compare to margined operation of the EDAC silicon, which then includes clock, detection window and short path padding overhead related to EDAC. Two-phase latch-based implementations such as [4, 6] benefit from having no short path padding requirements, but result in a significant overhead when transforming a flip-flop based design to a latch-based design. Pulsed latch-based designs as in [14, 15] have less overhead since they do not equip the slave latch, but often rely on a carefully controlled clock tree and duty cycle. A baseline pulsed latch implementation without EDAC system has the same short path padding constraints as its EDAC implementation, but requires significantly more short path padding than its flip-flop based counterpart. To the authors' best knowledge, this work is the first flip-flop based EDAC implementation fully functional at near-threshold supply voltages. Variation-resilient operation at these supply voltages is possible and was the goal of this work.

6.7 Conclusion

In this chapter, a lightweight timing error detection and correction strategy was presented. It is fully functional at ultra-low supply voltages down to 290 mV. The strategy augments existing architectures with timing error flip-flops, an error processor and a closed loop voltage control system to operate at the point-of-first-failure. The timing error flip-flop consists of a soft-edge flip-flop equipped with a transition detector to detect data arriving after the clock edge. As such, it detects 'real' timing errors. Detected errors are propagated to the system level to control the supply voltage, while the soft-edge flip-flop prevents pipeline corruption through time borrowing.

The EDAC strategy is implemented in a 32-bit ARM Cortex-M0 microcontroller system using an augmented standard cell VLSI design flow. The different design considerations regarding detection window size, sparse flip-flop replacement and time borrowing capability are analysed. The presented results are specific to the microcontroller system of this work, but the same approach is valid for any other digital system.

The EDAC-enabled microcontroller system is fabricated and measurement results have been presented. Ultra-low energy consumption (11–18 pJ/cycle) is achieved for a wide frequency range (5–30 MHz). The system is compared to a previously presented baseline silicon implementations and other state-of-the-art EDAC implementations. This work excels in ultra-low voltage capability thanks to its variation-resilient building blocks and in situ approach. It overcomes most of the performance predictability issues associated with near-threshold operated ultra-low energy systems. When looking at future advanced CMOS technologies, intra-die variations are expected to increase, while supply voltages decrease. This work has proven to be a viable solution for such conditions. The conclusions presented in this chapter were published in [10–12].

References

1. Bowman, K.A., Tschanz, J.W., Kim, N.S., Lee, J.C., Wilkerson, C.B., Lu, S.L.L., Karnik, T., De, V.K.: Energy-efficient and metastability-immune resilient circuits for dynamic variation tolerance. IEEE J. Solid-State Circuits **44**(1), 49–63 (2009)
2. Bull, D., Das, S., Shivashankar, K., Dasika, G.S., Flautner, K., Blaauw, D.: A power-efficient 32 bit ARM processor using timing-error detection and correction for transient-error tolerance and adaptation to PVT variation. IEEE J. Solid-State Circuits **46**(1), 18–31 (2011)
3. Das, S., Tokunaga, C., Pant, S., Ma, W.H., Kalaiselvan, S., Lai, K., Bull, D.M., Blaauw, D.T.: RazorII: in situ error detection and correction for PVT and SER tolerance. IEEE J. Solid-State Circuits **44**(1), 32–48 (2009)
4. Fojtik, M., Fick, D., Kim, Y., Pinckney, N., Harris, D.M., Blaauw, D., Sylvester, D.: Bubble Razor: eliminating timing margins in an ARM cortex-M3 processor in 45 nm CMOS using architecturally independent error detection and correction. IEEE J. Solid-State Circuits **48**(1), 66–81 (2013)
5. Joshi, V., Blaauw, D., Sylvester, D.: Soft-edge flip-flops for improved timing yield: design and optimization. In: IEEE/ACM International Conference on Computer-Aided Design, pp. 667–673. IEEE, Piscataway (2007)
6. Kim, S., Seok, M.: Variation-tolerant, ultra-low-voltage microprocessor with a low-overhead, within-a-cycle in-situ timing-error detection and correction technique. IEEE J. Solid-State Circuits **50**(6), 1478–1490 (2015)
7. Kulkarni, J.P., Tokunaga, C., Aseron, P.A., Nguyen, T., Augustine, C., Tschanz, J.W., De, V.: A 409 GOPS/W adaptive and resilient domino register file in 22 nm tri-gate CMOS featuring in-situ timing margin and error detection for tolerance to within-die variation, voltage droop, temperature and aging. IEEE J. Solid-State Circuits **51**(1), 117–129 (2016)
8. Kwon, I., Kim, S., Fick, D., Kim, M., Chen, Y.P., Sylvester, D.: Razor-lite: a light-weight register for error detection by observing virtual supply rails. IEEE J. Solid-State Circuits **49**(9), 2054–2066 (2014)
9. Partovi, H., Burd, R., Salim, U., Weber, F., DiGregorio, L., Draper, D.: Flow-through latch and edge-triggered flip-flop hybrid elements. In: IEEE International Solid-State Circuits Conference Digest of Technical Papers (ISSCC), pp. 138–139. IEEE, San Francisco (1996)
10. Reyserhove, H., Dehaene, W.: Design margin elimination in a near-threshold timing error masking-aware 32-bit ARM Cortex M0 in 40nm CMOS. In: 43rd IEEE European Solid-State Circuits Conference (ESSCIRC), pp. 155–158. IEEE, Piscataway (2017)
11. Reyserhove, H., Dehaene, W.: Design margin elimination through robust timing error detection at ultra-low voltage. In: IEEE SOI-3D-Subthreshold Microelectronics Technology Unified Conference (S3S), pp. 1–3. IEEE, Piscataway (2017)

12. Reyserhove, H., Dehaene, W.: Margin elimination through timing error detection in a near-threshold enabled 32-bit microcontroller in 40-nm CMOS. IEEE J. Solid-State Circuits **53**(7), 2101–2113 (2018)
13. Weicker, R.P.: Dhrystone benchmark (version 2.1) (1988). http://groups.google.com/group/comp.arch/browse_thread/thread/b285e89dfc1881d3/068
14. Whatmough, P.N., Das, S., Bull, D.M.: A low-power 1-GHz razor FIR accelerator with time-borrow tracking pipeline and approximate error correction in 65-nm CMOS. IEEE J. Solid-State Circuits **49**(1), 84–94 (2014)
15. Zhang, Y., Khayatzadeh, M., Yang, K., Saligane, M., Pinckney, N., Alioto, M., Blaauw, D., Sylvester, D.: iRazor: current-based error detection and correction scheme for PVT variation in 40-nm ARM Cortex-R4 processor. IEEE J. Solid-State Circuits **53**(2), 619–631 (2018)

Chapter 7
Conclusion

Abstract The focus of this work has been on realizing ultra-low energy consumption in digital systems. Implementation in CMOS technology is the most cost-efficient solution. A digital system implemented in CMOS technology typically shows a minimum energy point at near-threshold supply voltages. Ultra-low energy consumption is realized by implementing building blocks, architectures and design techniques that facilitate efficient near-threshold operation. In doing so, this work tackled a few challenges, thereby advancing the field of ultra-low energy digital systems and variation-resilient design techniques.

To start, functionality of digital building blocks at near-threshold supply voltages cannot readily be achieved. Transistors in sub-micron CMOS technology demonstrate significantly different behaviour at these voltages. The conventional logic building blocks result in sub-optimal performance. Moreover, speed performance degrades heavily. To serve most applications, MHz-range clock frequencies must be achieved. Additionally, transistors in sub-micron CMOS technology have variable performance due to, among others, process variations. This has to be taken into consideration when designing digital building blocks. Only few state-of-the-art implementations in literature demonstrate the combination of near-threshold functionality with a high speed and a high variation-resilience. Designing variation-resilient building blocks only gets you so far. Near-threshold operation will result in higher variation sensitivity. A highly variable system is not possible in most applications. The conventional approach to overcome this is accounting for the worst case, thus operating at a—for most conditions—sub-optimal point. Because near-threshold operation is so susceptible to these conditions, this worst case operation undermines all the initial effort to reduce energy consumption. The goal is an adaptive system that monitors the in-field variation condition of the system and operates it as lean as possible for this condition. This so-called point-of-first-failure operation has been demonstrated in literature through error detection and correction systems. Up till now, this has barely been adopted in near-threshold digital systems. The challenges that near-threshold operation brings require different considerations to optimally benefit from such a system.

Near-threshold variation-resilient building blocks that operate at the point-of-first-failure are not compatible with the conventional standard cell design flow.

H. Reyserhove, W. Dehaene, *Efficient Design of Variation-Resilient Ultra-Low Energy Digital Processors*, https://doi.org/10.1007/978-3-030-12485-4_7

Implementing an efficient design flow that can leverage the good properties of these techniques to very-large-scale-integration is mandatory. This can speed up the design time and manage the complexity of the large digital systems today. Moreover, multi-supply voltage timing and power analysis can enable minimum energy considerations in terms of static vs. dynamic energy at design time.

To demonstrate the applicability and low energy results of these techniques, a proof-of-concept system is necessary. Microcontrollers are wide-spread digital systems that can benefit from ultra-low energy consumption. The ultra-low energy implementation of such a microcontroller system faces all these challenges and acts as a proof-of-concept, preferably with better than state-of-the-art performance.

This chapter reconsiders the previous chapters and their conclusions. In doing so, it demonstrates how these challenges were overcome throughout this work.

7.1 General Conclusions

Chapter 2 focused on enabling near-threshold operation in CMOS technology. By looking at the most basic transistor properties, the challenges that normally prevent near-threshold operation became clear. It was shown that the key to realizing variation-resilient near-threshold operation is maximizing the noise margin. This is done by balancing the pull-up pull-down network through ratioed logic. The inverter is implemented with a stacked nMOS pull-down network. It is demonstrated that this overcomes the excessive pMOS-nMOS ratios that near-threshold operation requires and improves the imbalance in the leakage–drive current ratio. Moreover, it reduces the total area and the input capacitance. Exploring logic functions, the choice to implement differential transmission gate logic was made and motivated. The identical pull-up pull-down network is inherently balanced, and the pMOS-nMOS pair enables variation-resilient operation. By using LVT devices in a general-purpose technology flavour, a high-speed performance is achieved. This is clearly demonstrated through a study with different V_T devices. The inverter and the transmission gate are combined to make up every logic function, including latches and flip-flops. These are the logic gates and flip-flops that make up the high-speed variation-resilient digital systems in this work.

Chapter 3 composed an efficient VLSI design flow. By looking at the typical VLSI design flow, a few key challenges came up when thinking about near-threshold operation. Especially the use of differential transmission gate logic is not readily supported. The generation of a fully differential transmission gate standard cell library, including simulation, characterization and layout, was presented as the first step to variation-resilient near-threshold VLSI design. An adapted VLSI design flow was presented. It uses all the commercial EDA tools as normal, and augments or intervenes in some steps with a scripted approach. The final result was an effective standard cell design flow-library combination that allows efficient implementation from register-transfer description down to layout.

Chapter 4 demonstrated two ultra-low energy proof-of-concept implementations. It explored the architecture of the 32-bit ARM Cortex-M0 microcontroller and its framework to make a workable silicon implementation that runs in a lab test environment. The implementation details are presented, leading to a silicon implementation of each of the two prototypes in 40 nm CMOS technology. Measured results are presented of both. For the first prototype, these results demonstrated possibilities for improvements. The second prototype implemented these improvements, as is demonstrated in a comparison of measurement results of both. Speed performance and energy consumption across the entire near-threshold operating range are shown. By looking at state-of-the-art literature that implements similar prototypes, it was shown on what and how these prototypes improve: the combination of a high-speed and an ultra-low energy consumption.

Chapter 5 studied error detection and correction systems. Different strategies that improve performance are considered, to come to a generic system overview. It was shown that a wide range of architectures that demonstrate different considerations is available in the literature. Different detection strategies and their circuits can be combined with different correction strategies and their architectural implications. When operation in the near-threshold region is to be enabled, more challenges arise. These challenges were identified and elaborated on. The key consideration is to reduce total energy consumption by operation closer to the point-of-first-failure through dynamic voltage scaling, while taking into account that the error detection and correction mechanism also consumes energy. A qualitative analysis made clear what the different components are to consider when looking at the performance of an error detection and correction system. These components were discussed, to make clear if, how and on what a system improves when compared to a baseline system without error detection and correction that operates with worst case margins.

Chapter 6 put the error detection and correction considerations in practice by implementing a timing error-aware microcontroller system. A detailed overview of the proposed error detection and correction mechanism was given, as well as how it was integrated in the ARM Cortex-M0 microcontroller system. The timing error-aware flip-flop that was demonstrated uses a near-threshold enabled transition detector to detect late arriving data, and overcomes timing errors by masking them by means of time borrowing. In-flip-flop detection window generation results in careful control of the detection window and the allowed time borrowing amount. Through an overview of the circuits and simulations of this technique, its functional properties became clear. Automatically integrating error detection in the VLSI design flow is non-trivial and was covered in full extent. To minimize overhead, a sparse flip-flop replacement was implemented relying on logic path propagation delay statistics, and a small but reliable detection window was chosen. Finally, the measured results of the system implemented in 40 nm CMOS technology were presented. A state-of-the-art comparison puts these results in perspective. The key improvements are shown to be operation at near-threshold supply voltages, lightweight error correction and well-considered sparse flip-flop replacement and timing detection window generation and size which minimizes overhead.

Through these five chapters, this work built up unique insights into the minimum energy operation of digital systems. Considerations were presented on the architectural level, passing through the design flow, all the way down to the transistors that make up the standard cell libraries. By going through the entire top-down design process, it was possible to come to conclusions that would have not been possible otherwise. These insights were achieved by starting at the transistor level, gradually increasing up the hierarchy by means of different abstraction levels, adding complexity, to end at the final prototype that incorporates all of the above: **an ultra-low energy consumption, high speed, variation-resilient microprocessor that operates at its point-of-first-failure autonomously, implemented in CMOS technology**. The following more general insights were established:

- Ultra-low energy consumption does not equal low performance. A combination of the correct transistor type choice, circuit techniques and architectural considerations results in a MHz-range speed. At these speeds, major energy savings can be realized.
- The wide adoption of conventional static CMOS logic does not necessarily means it is the best choice. Especially when near-threshold operation results in a large asymmetry between pMOS and nMOS devices, as is the case in the technology in this work, it does not perform optimal.
- The typical VLSI design flow is efficient, but should not always be adopted as is. Automated adaptations, interventions and augmentations are possible and can improve its performance and its variation-resilience.
- Standard cell libraries are usually provided by the technology foundry. However, the design of a custom (sub-) set of standard cells is possible with reasonable effort and careful simulation. Re-characterizing an existing standard cell library for ultra-low voltage operation is another possibility.
- Variation-resilience is a full package deal. Variation-aware decisions are necessary at every step of the design process: in the logic building blocks, the design flow and the architecture. Only through a combination of techniques can the best possible result be achieved.
- The minimum energy point is the design target, but the optimum is quite flat. This means low energy consumption can be achieved in a relatively wide range around the minimum energy point without too much compromise.
- Near-threshold operation of a timing error detection and correction system is possible. Variation sensitivity makes it hard, but near-threshold operation has as much to gain: the worst case margins that are overcome increase with the sensitivity. If the right architecture is combined with efficient circuits, significant improvements can be made.

The insights presented in this work were presented at a selection of international conferences, and published in full in conference proceedings and journals. The following publications are the result of this peer-reviewed publication process: [3–8].

7.2 Future Work

This work has advanced the field of ultra-low energy digital systems and variation-resilient design techniques by gaining a clear understanding of the considerations in these fields and developing prototypes that improve thereon. As such, it is an important step towards the adoption of near-threshold systems in applications. Nevertheless, it left some parts towards this goal untouched. Several improvements and other research directions are still possible. In net energy consumption, memory optimization and standby power optimization can bring the most benefit. Statistical static timing analysis is the most constructive approach to improve insights and benefit from them.

- **CMOS technology** is getting more and more advanced. More advanced nodes than the technology used in this work can improve near-threshold operation, and more generally, ultra-low energy operation. Although the effects mentioned in Chap. 2 regarding pMOS-nMOS imbalance are still present, they are becoming less pronounced in these technologies. Moreover, more exotic technologies like FD-SOI technology, also discussed in Chap. 2, demonstrate extremely good properties for ultra-low energy implementation. The same goes for the even more advanced finFET technology. A steeper sub-threshold slope can yield very good near-threshold properties. It is clear that these technologies bring significant advancements in both speed and energy consumption to the systems discussed in this research. It remains to see whether the applications mentioned in Chap. 1 can be fabricated in a cost-effective way in these technologies.
- **Operation down to the minimum energy point** is the most energy-efficient way for a digital system to function. All the prototypes developed in this work are functional at supply voltages well below the minimum energy point. Enabling operation there comes at a cost. Ideally, logic functionality should be realized down to the minimum energy point, **but not further**. Creating a logic library optimized for the target minimum energy point supply voltage is the best choice. The MEP is architecture and activity sensitive. Thorough insights in the application workload should coincide with minimum energy considerations.
- **Statistical static timing analysis** is a recent innovation in the VLSI design flow that can reduce the worst case margins, as discussed in Chap. 3. To the authors best knowledge, foundries are only providing models or standard cell libraries that enable SSTA for extremely variation sensitive technologies well below 40 nm CMOS. However, ultra-low voltage operation is extremely variation sensitive as well and can thus benefit in a similar way. Developing a new standard cell library that enables SSTA or re-characterizing the existing libraries to this end will result in better insights. Especially the considerations regarding sparse EDAC flip-flop replacement can be extended significantly with automated statistical analysis. It will enable easier and faster design considerations in this regard and will reduce the EDAC energy overhead even further.
- **Ultra-low voltage clock distribution is hard.** It typically results in a large insertion delay. For bigger systems than the ones discussed in this work, this may compromise reliable clock distribution. Pu et al. [2] solve this by implementing

a network-on-chip (NoC) to communicate between the larger sub-blocks of the system. This overcomes the need for a synchronized clock between all of them. For far bigger systems than the ones discussed in this work, such a NoC may provide a solution. The impact of such an approach is expected to be quite big, but is necessary for functionality and performance.

- **Efficient testability of digital systems** is a necessity for industrial adoption. This work did not explore design-for-testability techniques particular for near-threshold operation nor did it implement testability in the prototypes by means of scan flip-flops. Although this was not investigated further, no insurmountable challenges are expected in doing so.
- **The energy efficiency of the microcontroller system** was not explored to its fullest. A major improvement can be realized by equipping it with a low energy, MHz-speed optimized SRAM memory. Moreover, architectural techniques, such as clock gating, fine-grained power domain gating, etc., are not implemented. Improving the standby power of digital systems was not one of the main goals of this work. In some applications, the system may be idle for longer periods of time. In this case, a large standby power is not advised. Most of the state-of-the-art techniques to decrease standby power could be applied in this work as well. These could improve the energy consumption even further. These two fields are currently the main components that prevent this research from being applicable in an industrial context. A reasonable amount of work and well-targeted memory IP can go far in this regard.
- **Evaluation of variation-resilience** on fabricated silicon is hard work in a research environment. To evaluate the variation-resilient properties of some of the techniques presented in this work to the fullest, multiple measurements of a large number of dice, fabricated in different process conditions, are necessary. These analyses are typically not done in academic approaches such as this work, but should demonstrate the variation-resilience of the demonstrated techniques.
- **The dynamic voltage scaling control loop** as used in this work relies on an off-chip power management unit. The integration of a DC/DC converter with the presented prototype on a single chip has not been realized. Doing so will demonstrate the effectiveness of such an approach, as well as the trade-offs it presents regarding DC/DC conversion efficiency vs. microcontroller efficiency.
- **Compensating other than process variations** with the error detection and correction system is possible, but may shift some of the trade-offs presented in this work. Especially when timing errors are triggered by fast-changing variations such as voltage-droop, correction through error masking may not function properly. A careful study of it, when and where these variations occur, followed by improving the error detection and correction system or looking at other techniques to overcome these variations should follow.
- **Short path padding because of increased hold time** is the main drawback in the presented error detection and correction mechanism. The impact of short path padding depends entirely on the architecture and its combination of critical and short paths. In some architectures, this may compromise the good properties of error detection and correction. Other approaches should be applied. One option

is to create a hybrid system that uses in situ critical path *replicas* that are continuously active to do error detection on. This overcomes the need for short path padding, but adds overhead elsewhere. Short path padding is so architecture sensitive that it needs to be evaluated in a case-by-case manner.

- **Other variation-resilient design techniques** have recently been demonstrated that may improve energy consumption even further. Gangopadhyay et al. [1] and Sun et al. [9] demonstrate unified voltage and frequency regulation techniques (UVFR). The key idea here is a unified approach where the delay-locked loop and the voltage regulator are combined into a single building block. Because of this unified approach, most of the variation effects are shared between voltage and frequency. In this way, the speed performance and the supply voltage always track each other. These techniques reduce voltage margins and could probably be applied for near-threshold operation as well. The system level considerations of UVFR are not to be underestimated and could prevent applicability of such techniques in most systems.

References

1. Gangopadhyay, S., Nasir, S.B., Subramanian, A., Sathe, V., Raychowdhury, A.: UVFR: a unified voltage and frequency regulator with 500MHz/0.84V to100KHz/0.27V operating range, 99.4% current efficiency and 27% supply guardband reduction. In: 42nd IEEE European Solid-State Circuits Conference (ESSCIRC), pp. 321–324. IEEE, Piscataway (2016)
2. Pu, Y., Shi, C., Samson, G., Park, D., Easton, K., Beraha, R., Newham, A., Lin, M., Rangan, V., Chatha, K., Butterfield, D., Attar, R.: A 9-mm2 ultra-low-power highly integrated 28-nm CMOS SoC for internet of things. IEEE J. Solid-State Circuits **53**(3), 936–948 (2018)
3. Reyserhove, H., Dehaene, W.: A 16.07pJ/cycle 31MHz fully differential transmission gate logic ARM Cortex M0 core in 40nm CMOS. In: 42nd IEEE European Solid-State Circuits Conference (ESSCIRC), pp. 257–260. IEEE, Piscataway (2016)
4. Reyserhove, H., Dehaene, W.: A differential transmission gate design flow for minimum energy Sub-10-pJ/Cycle ARM Cortex-M0 MCUs. IEEE J. Solid-State Circuits **52**(7), 1904–1914 (2017)
5. Reyserhove, H., Dehaene, W.: Design margin elimination in a near-threshold timing error masking-aware 32-bit ARM Cortex M0 in 40nm CMOS. In: 43rd IEEE European Solid-State Circuits Conference (ESSCIRC), pp. 155–158. IEEE, Piscataway (2017)
6. Reyserhove, H., Dehaene, W.: Design margin elimination through robust timing error detection at ultra-low voltage. In: IEEE SOI-3D-Subthreshold Microelectronics Technology Unified Conference (S3S), pp. 1–3. IEEE (2017)
7. Reyserhove, H., Dehaene, W.: Margin elimination through timing error detection in a near-threshold enabled 32-bit microcontroller in 40-nm CMOS. IEEE J. Solid-State Circuits **53**, 2101–2113 (2018)
8. Reyserhove, H., Reynders, N., Dehaene, W.: Ultra-low voltage datapath blocks in 28nm UTBB FD-SOI. In: IEEE Asian Solid-State Circuits Conference (A-SSCC), pp. 49–52. IEEE, Piscataway (2014)
9. Sun, X., Kim, S., ur Rahman, F., Pamula, V.R., Li, X., John, N., Sathe, V.S.: A combined all-digital PLL-buck slack regulation system with autonomous CCM/DCM transition control and 82% average voltage-margin reduction in a 0.6-to-1.0V cortex-M0 processor. In: IEEE International Solid-State Circuits Conference Digest of Technical Papers (ISSCC), pp. 302–304. IEEE, Piscataway (2018)

Index

H. Reyserhove, W. Dehaene, *Efficient Design of Variation-Resilient Ultra-Low Energy Digital Processors*, https://doi.org/10.1007/978-3-030-12485-4

Printed in the United States
By Bookmasters